IS THE ALGORITHM PLOTTING AGAINST US?

IS THE
ALGORITHM

PLOTTING
AGAINST US?

A Layperson's Guide to the
Concepts, Math, and Pitfalls of AI

KENNETH WENGER

WORKINGFIRES
F O U N D A T I O N

New York

Cover design by Fayyaz Ahmed
Cover image used under license from Shutterstock.com
Original book design and layout by James Protano

Library of Congress Control Number: 2022947822
ISBN 978-1-959632-01-6 (paperback)

Working Fires Foundation
1879 Whitehaven Road, Ste. 2040
Grand Island, NY 14072

workingfires.org

To my family

With regard to the question of whether we can make [machines] think like [human beings], my opinion is based on the following idea: that we try to make these things work as efficiently as we can with the materials that we have. Materials are different than nerves and so on. If we would like to make something that runs rapidly over the ground, then we could watch a cheetah running, and we could try to make a machine that runs like a cheetah. But it's easier to make a machine with wheels, with fast wheels, or something that flies just above the ground in the air. When we make [an airplane], the airplane doesn't fly like a bird. They fly, but they don't fly like a bird. They don't flap their wings exactly. They have in front another gadget that goes around, or the more modern airplane has a tube that you heat the air and squirt it out the back—jet propulsion. A jet engine has internal rotating fans and so on and uses gasoline. It's different, right? So there's no question that the later machines are not going to think like people think, in that sense. With regard to intelligence, I think it's exactly the same way. For example, they're not going to do arithmetic the same as we do arithmetic, but they'll do it better.

—Richard Feynman, "Computers from the Inside Out" (1985)

CONTENTS

ILLUSTRATIONS

Figures

Tables

IS THE ALGORITHM
PLOTTING AGAINST US?

INTRODUCTION:
LIVING WITH LIONS

S ome days it feels like the whole world just can't stop talking about artificial intelligence, or AI. Some of it seems good and exciting, like self-driving cars. We can already see cars maneuvering in certain situations with little human interaction; it won't be long before the driving experience is all but automated. Some of it seems straight out of *Star Trek* or an Arthur C. Clarke novel. Neuralink, a company cofounded by Elon Musk, promises futuristic chips that can be inserted into your brain and interface with your neural connections, initially to help persons with disabilities gain lost functionality and eventually to serve as a much faster interface with our digital world. Imagine surfing the web (Does anyone still say "surfing the web"?) without needing a keyboard. Then, there are voice assistants like Amazon's Alexa and Apple's Siri that can understand our endless queries and respond in impressive ways. All these advances might have you wondering how any of this is possible. What is the source of knowledge in these machines, and how do they actually work?

Darker and more ominous undercurrents, however, have also sparked interest in AI. If you haven't watched the Netflix series *Black Mirror*, you should. It's great, and it's also a reminder—or maybe a warning—of what can happen when we lose control of our technology. Maybe you're reading this book to find out exactly how terrified you should be of AI. Maybe you've read news articles or seen Instagram videos about companies likes Boston Dynamics creating animallike robots that can maneuver in complex environments, perform such sophisticated tasks as opening doors, and communicate with other robots to achieve a common goal. Maybe you've thought, "Oh my God, it's too late. We are all going to die!"

Whether your interest in AI is driven by hope and excitement or gloom and despair, you want to know if this book is for you and what you will get out of reading it. So let's get right to it.

The purpose of this book is, first and foremost, to explain how AI works at a level of detail that makes these algorithms accessible to a general audience. You do not need a technical background to understand this book; all that is required is a sense of curiosity and a willingness to consider complex subjects. After reading the book, you should have a good handle on the capabilities of state-of-the-art AI algorithms so that you can evaluate and gauge your response to this technology from an informed place.

A lion in the wild can be dangerous to humans. But knowing where they live, when they hunt, and their physiological capabilities can help us modulate our response and behavior. If we find ourselves on a safari in the African savanna, we should be alert. If we see a lion, we might want to stay in our vehicles and keep enough distance so that we can drive away if the lion decides to chase. We can do this because we know how fast our vehicle can travel, and we know the speed of a lion. We also know that lions can run, but they can't fly. This knowledge helps us gauge the level of readiness we ought to have in the African savanna. The Maasai of Tanzania and Kenya understand this better than anyone. They have coexisted with lions for millennia. They use knowledge developed over generations to keep themselves and their herds safe from lions. Once, long ago, they used their skills to track and successfully hunt lions in traditional rites of passage. Now, increasingly, they use those same tracking skills to protect and help preserve the king of the jungle. The Maasai do not fear the lion; instead, they have learned to understand it.

On the other hand, if we find ourselves in the jungles of South America, we know that we need not fear lions—other predators, sure, but not lions, because there are no lions outside of Africa. In other words, understanding the lion's capabilities, limitations, and domain helps us understand when we must worry about lions and when we can be certain that we are safe from them. That is the goal of this book (not to discuss lions—though we will come back to them in a later discussion!): to help us understand the capabilities, limitations, and domains of current AI technology.

First, we need to define what we mean by AI in this book. When we say "AI," we are referring to a specific class of AI algorithms: artificial neural networks. Readers may have heard of *deep learning* used in conjunction with AI these days as well. This term often describes neural network models with multiple layers of artificial neurons. We come back to this relationship in chapter 1 and discuss the significance of each layer in a neural network model. It is important to note, however, that AI is a broad discipline in computer science; it spans many areas of research, and countless algorithms fall under this umbrella term. We are aware of the umbrage taken by purists—and those inclined to proper definitions, terminology, correctness, and so on—when we use the terms *AI* and *neural networks* interchangeably in this book, but the reality is that most people ignore this distinction often enough that those terms are regularly used interchangeably in informal contexts. Before proceeding, let's make a promise to eternally remember that AI is a broad discipline, and neural networks are a class of algorithms belonging to that discipline. Having promised to remember the distinction, we can now do as we please.

Artificial neural networks are by far the most popular and successful AI algorithm in use today. They are currently the driving force behind advances in robotics, self-driving cars, Amazon's Alexa, and Google Assistant. In the pharmaceutical industry, it is expected that neural networks will make significant contributions to the discovery of molecules that can be synthesized to treat debilitating diseases like Parkinson's, multiple sclerosis, and many others. At this point, it seems like there is no problem big enough or abstract enough that artificial neural networks cannot handle it. The corollary to this is that their unprecedented level of success has also garnered for them a certain level of distrust, or at least trepidation. If it can respond to me and understand my queries so well, what else can it do? What is it thinking?

Popular beliefs, science fiction, and media sensationalism have conditioned us to distrust AI systems. That general distrust has taken on one most prominent and particular flavor: these machines will eventually become self-aware, wake from their eternal slumber, and kill us. The problem with such concerns is that they serve as a bit of a red herring. Regardless of whether we will eventually have to contend with self-aware machines, it's certainly not the only issue we should be discussing at this time. Presently, science does not have a complete theory of consciousness. We do not understand how consciousness arose in ourselves. We don't even have a definition of consciousness that everyone agrees with. Recent advances in AI—mixed with a lack of understanding of how AI systems work and a tendency on the part of media companies to generate revenue by stoking fears—contribute to a general sense that conscious artificial systems are just around the corner, ready to enslave us. Society may have to grapple with conscious AI in a distant future, but before we get there, plenty of more urgent matters warrant discussion: What happens when AI is used in ad campaigns? What about by law enforcement? Can AI algorithms solve a problem in unique ways, where its measure of "success" differs from that of its human designer? This last example is the well-known *alignment problem*. Suppose you ask a robot to get rid of the CO_2 in the atmosphere to combat climate change, and it finds that the best way to achieve this goal is to get rid of the human population. The robot did not consciously decide to get rid of humans. The situation simply describes an optimization problem gone wrong (for us!). Understanding when our current AI technology is being applied and the potential unintended consequences of its misuse are real-world, present-day issues that get pushed aside because they are not as exciting as the thought of berserk smart blenders chasing us around the house.

In this book, we first address the functioning of neural network algorithms. We explain what it is that makes them tick and how they manage to work at all. Then, we critically examine their limitations, the rights and trust we have already granted them, and their potential for causing significant harm to our society, in some cases, if we are not careful. But let's be clear: this harm is completely self-inflicted. The algorithms are not yet "out to get us"; we just don't always use them in healthy and productive ways.

Why should you get involved in this discussion if you are not a scientist? Because each of us can influence our collective future. Technology advances with research, and research is fueled by money. Institutions get billions of dollars in government grants for research into different areas. The grants are made possible by taxpayer money—your tax money. You have the ability to influence policy every time you go out and vote. The public can decide what areas of research should get more attention. But how can you make an informed decision without being, well, informed? When it comes to artificial intelligence, there is a lot of speculation, often in the media, about the dangers and the capabilities of AI. You will be much better served by understanding how these systems work and what we realistically need to worry about rather than making an emotional decision based on uninformed sensationalist ideas. This way, you can at least arrive at a decision by way of a thought process. If we don't have a thought process to ground our decision-making, as often happens, we get disproportionate—typically radical and extreme—responses driven by fear. This happened when stem cell research was all but banned in many countries out of fears over possible misuse: critics concocted specious moralistic arguments and completely disregarded the ethical dilemma of abandoning research that could contribute significant insight into terrible diseases like cancer, AIDS, and degenerative muscle disorders.

For many of us, AI immediately conjures up the specter of Skynet—an intelligence created for the purpose of protecting national security that inevitably gains consciousness and wreaks havoc on humanity—and its cyborg assassin, the Terminator T-800. But our fear of AI might derive from a more primitive and innate response to a perceived threat, a response that predates the development of technologies whose imagined descendants populate movies and science fiction novels—that is, the fear of the unknown. More specifically, that fear has often manifested as a fear of the Other: a creeping, gut-borne feeling characterized by increasing and alarming suspicions of newcomers, outsiders, or anyone beyond our circles of intimacy or relationality. What are these circles? Interestingly, we construct different ones depending on certain rules of engagement. First, there is the family circle, where we extend the most trust. Beyond this, we have friends and more distant relatives. Even our respective countries form a certain circle of trust, if not comfort; we typically feel more connected to our compatriots than to people from other parts of the world. We notice this when we travel and meet a fellow expat. Immediately we feel a connection to them even though we know very little about them; we just know that they belong inside one of our circles.

In its beneficial, or at least benign, form, this distinction between insiders and outsiders can foster a sense of community. In its malignant form, as the twentieth century showed in unconstrained horror, it leads to xenophobia and fascism (and unfortunately such virulent nationalism has been on the rise again around the world). So far, we are just talking about relationships between people. What then can we expect from our relationships with other beings, including the artificial kind? It seems natural that we should be suspicious of artificial intelligence. In some ways, it is the ultimate threat: by definition an outsider, not being human or natural, yet possessing the crown jewel of all qualities that separate us from mere animals—intelligence.[1] Intelligence has given our species the superpower to change our planet and dominate all other living things on it. When looking at the full scope of what we have done with our intelligence—taking in our remarkable creations in the arts and sciences, our developments and advances living as social beings—in some very specific and clear ways, it has not served the natural world well: destroying forests, polluting oceans and waterways, wiping out entire species of plants and animals, and threatening those that are still around. It is no wonder that we should be outright mortified of a being, or entity, that shares very little with us yet finds itself in possession of our ultimate weapon.

It seems that our primary concern, then, is with the form and extent of so-called intelligence in artificial systems. Over the course of this book, we systematically describe the mechanisms that are responsible for the advances we see today. Once we finally understand the machinery behind AI, we should feel empowered to take charge of our technology instead of being fearful of some synthetic omnipotence. At the very least, establishing a foundational understanding will give us the tools to judge what we should accept and what we should constrain when it comes to artificial systems that are now making decisions that affect our lives. My hope is that this book will not leave you feeling afraid but rather informed and, therefore, empowered.

---

1. Apes, dolphins, and even some species of birds like crows and magpies are known to be quite intelligent. When it comes to human-level intelligence, however, it's quite clear that we stand alone. In the context of discussing possible alien civilizations, the scientist and educator Neil deGrasse Tyson has postulated that the difference in DNA between chimps and us is 1 percent on average. He asks the question, "If the difference between humans and chimps is driven by that one percent, and that one percent is responsible for an 'intelligent' chimp stacking boxes and an intelligent human building the Hubble space telescope, what might the difference be between an advanced alien civilization and us, even if they are just one percent smarter than us?" Now consider an AI that is just a few percent smarter than we are. "Neil deGrasse Tyson: Only 1% Separates Our Intelligence from Chimps," YouTube video posted by Danica Patrick on Sept. 6, 2019, 4:54, https://www.youtube.com/watch?v=F200wpEpJ4w.

○ ● ○

The rest of the book comprises four main chapters and a short conclusion. In chapter 1, we discuss the first artificial neuron created by humans and the history and motivation that led to its creation. Along the way, we meet individuals who were driven purely by curiosity—that insatiable need to understand everything about our universe, from its physical laws to the phenotypic expressions of those laws. We examine how the humble artificial neuron evolved into a network of artificial neurons powerful enough to solve problems that were once considered computationally intractable, such as image recognition and natural language processing (e.g., understanding speech and writing, translating between languages). Wherever possible, we point out similarities between artificial and biological neurons, similarities that inspired the early work in artificial intelligence. Importantly, we unpack the multilayer perceptron—the first artificial neural network architecture ever created and still a fundamental building block of most state-of-the-art architectures in use today.

Chapter 2 is all about vision. Here we discuss the problem that computer vision presents. Today, we take for granted that cameras and gadgets can track our faces and follow our movements. Even relatively inexpensive drones can be programmed to track and film us as we ski down a mountain. We search images for content using a variety of applications and tools. These tools can take a search query from us (something like "pictures of red cars in autumn"), analyze images for content (that shows red cars in autumn), and then return a list of images matching these search criteria. In the medical domain, similar applications are capable of searching biopsy scans for anomalous tissues that might be signs of disease. How is any of this possible? If you keep up with technology, none of this is surprising or even impressive anymore. But computer vision was once considered among the most elusive subjects to tackle in computer science. In chapter 2, we see why computer vision is such a fundamentally difficult problem to approach, and we discuss how artificial neural networks have all but solved this problem. And finally, we introduce the convolutional neural network, which has become the de facto architecture for computer vision. In fact, together with the multilayer perceptron, they form a set of fundamental building blocks used in most neural network architectures today. Throughout, we continue to note the similarities between artificial and biological systems and, wherever possible, describe the biological system as the inspiration for and intuition behind the development of the artificial one.

The overarching goal of the first two chapters, then, is to specify what neural networks look like and how they operate. Providing the layout of the neural networks (in graphic and linguistic form), we describe each layer and show how each neuron in one layer is connected to the neurons in subsequent layers. We also detail the types of operations taking place at each neuron. After reading chapters 1 and 2, you should be able to answer—at least at a conversational level—what neural networks are and how information is processed from the input to the output.

In chapter 3, we dive deeper into the how questions. This is where we look at what gives artificial neural networks any right to work and elucidate the mathematical intuitions that govern most of them. We describe the training processes that enable neural networks to perform certain tasks. This information allows us to understand their limitations and start to grasp the current state of artificial intelligence. We explore whether these systems are capable of conscious thought—whatever that means to you—or whether they are enacting a more primitive method of information processing.

In the final main chapter, we take a step back from our pursuit of understanding neural networks in specific and operational terms. Instead, we put to good use the information we learned in the preceding chapters and attempt a bit of introspection. Using our newfound knowledge, we again ask the question of how dangerous artificial intelligence is and proceed to answer the question by evaluating levels of threat. We discuss areas (the judicial system, advertising) where artificial intelligence poses significant risk—without requiring the technological leap of gaining consciousness—and examine industries (automotive, health care, warehousing) where automation promises an improvement over present-day standards. More importantly, in this chapter we evaluate the current AI revolution against previous technological revolutions and attempt to learn from the past to understand our current moral and practical obligations.

Having defused the panic about a robot takeover, in the conclusion, we provide a simple test for identifying classes of problems that are amenable to AI-based solutions and classes of problems that should remain in human control for the foreseeable future. The test provides a path to action by asking a set of questions for any new class of problems we may want to solve using automation. This set of questions enables us to reflect on the problem to understand whether the solution requires nuanced and difficult moral considerations or simply a set of specific rules to follow. Armed with these simple questions, we can then take control of deploying our AI tools in responsible ways.

So let's get to it. The path to learning whether Alexa is conspiring with Siri begins with chapter 1—including a brief diversion into what amounts to present-day technology's ancient past.

1

POLARIZATION
AND ITS
CONSEQUENCES

What are artificial neural networks? Are they really made of neurons, like our brains? The structure of artificial neural networks presents a starting point in our pursuit to understand them but says very little about their capabilities and limitations. For this, we must dig deeper and ask more pointed questions. Why were they invented, and what kinds of problems can they solve? How do they know things? Can they learn on their own, or do we have to teach them? In this chapter, we dedicate considerable time to understanding artificial neural networks and their history. Their origin story is one of optimism and hope.

As you might imagine, the artificial neural network, much like our own brains, began its life as a single cell. Complex networks developed from that cell, and these networks are today's rock stars of artificial intelligence. We use them for making sense of data. We use them to classify images, performing object recognition and tracking for self-driving cars. You might have seen a Tesla successfully merge from one lane to another without human intervention. The car has cameras and other sensors that help it capture the state of the outside world, but artificial neural networks are tasked with interpreting the data from its sensors to identify vehicles, pedestrians, traffic signs, road signs, and anything else that enables it to *see* the world. Researchers are currently investigating using neural networks in medical settings as image classifiers to help diagnose different kinds of diseases, from melanoma to breast cancer to pulmonary ailments like pneumonia. For example, neural networks may be tasked with identifying anomalous regions in image scans of biopsied tissue or signs of pulmonary effusion in X-ray images. In a more general sense, we use artificial neural networks to process large volumes of data and extract patterns and trends from the data that might be significant to us. In some cases, this involves analyzing images, but neural networks are not limited to vision applications. We might train a neural network to predict the price of houses in certain neighborhoods or teach it to analyze financial markets' historical data to forecast future trends. In other words, we use artificial neural networks as tools to solve certain classes of problems— namely, *classification problems* and *forecasting problems*.

But the history of artificial neural networks, in some ways, is not the pursuit of a tool for artificial intelligence. It started with researchers' need to understand the human brain. Researchers like to use analogs and models to investigate concepts. It is difficult, however, to tinker with a real brain; people get in the way. Early researchers in neuroscience and psychology thought it would be useful to try to build an artificial brain as a step to understanding our own.

To try to answer the questions we opened with, this chapter is broken into a few sections. We start by taking a stroll through the halls of history and examining the earliest steps (in the nineteenth century!) that contributed to building a functioning artificial neural network. We discuss how researchers came to view the neuron as the key element in information processing and look at the first artificial neuron: the McCulloch-Pitts neuron. We explain how developments in our understanding of the biological brain contributed to modifying the McCulloch-Pitts neuron into a more powerful system called the perceptron, which led to the simplest type of artificial neural network—the fully connected neural network, also known as the multilayer perceptron, or MLP (fig. 1.1). We explain why, although artificial neural network research began in earnest in the 1950s, you probably didn't hear about such networks until very recently: research in the field suffered many ups and downs from its inception, with many of the downs driven, in part, by overhyped promises. We try to make the case that, while hype is still a problem, it looks promising that artificial neural networks will stick around. Once we establish the progression of steps that led to modern artificial neurons, we examine how these neurons can be combined into complex networks. We look at examples and explain how information is processed by each neuron in the network and how the output of the network is calculated and interpreted. Finally, we close the chapter with a few use-case examples for how we can employ neural networks to solve real-world problems.

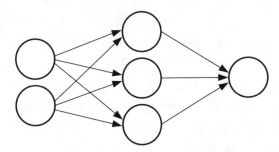

Figure 1.1 A simple artificial neural network. Information flows from left to right. The *two circles* in the first layer are the input nodes. The *three circles* in the middle are the processing neurons, and the *one circle* on the right represents the output value. The output in a neural network signifies a "prediction" on the input.

DISENTANGLING THE NEURON

Before we jump into the more technical aspects of neural networks and how they function, let's spend a few moments learning about their history. Understanding what early researchers did to decipher the mysteries of the brain will help shed light on why artificial neural networks operate the way they do.

To apprehend neural networks, biological or artificial, we must begin with the neuron. We now know that our brains are made of billions of neurons and that neurons are the fundamental building blocks of information processing. But we didn't always know this. And, as is typical of scientific advancements, the discovery of the neuron and interpretation of its function were not without contention. How did we come to view the humble neuron as the fundamental unit of processing? It starts with a young nineteenth-century Spaniard exuding energy and enthusiasm. His name was Santiago Ramón y Cajal. He was the first person to understand that the brain was made of individual neurons and that these neurons played a pivotal role in information processing.

Cajal was born in 1852 in a small village in Aragon, in northeast Spain. As a young man, Cajal's first inclination was to become an artist. He enjoyed drawing the natural world, a passion that would come in handy in his future scientific career (figs. 1.2–1.5). His father was a surgeon and professor of dissection at the University of Zaragoza. The younger Cajal eventually followed in his father's footsteps and enrolled in the Zaragoza school of medicine, graduating in 1873. Cajal's big contribution to neuroscience began in 1887. That year, Cajal traveled from Valencia to Madrid to learn about new technological advances related to sample preparation for inspection under a microscope. There he met a brilliant psychiatrist by the name of Luis Simarro Lacabra, who showed him brain specimens stained using a technique developed fourteen years earlier by the Italian Camillo Golgi. The technique involved hardening a piece of brain matter in potassium dichromate and later dousing it with silver nitrate. This had the effect of dyeing only a few types of cells, revealing their complete structures as black silhouettes against the unstained background. Those who knew of this technique—not many, as the technique had not enjoyed great dissemination in the fourteen years since its inception—knew it as Golgi's *reazione nera* (black reaction). Upon seeing the specimens produced by Lacabra, Cajal quickly realized the inadequacies of the current methods for studying nervous tissue. He would later write in his autobiography that the staining technique produced cells "coloured brownish black even to their finest branchlets, standing out with unsurpassable clarity upon a transparent yellow background. All was sharp as a sketch with Chinese ink."[2] Cajal used and improved the staining technique as he studied the neural tissue composition of the retina, cerebellum, and spinal cord.

At the time, the prevailing scientific consensus, driven in large part by the German histologist Joseph von Gerlach, was that nervous tissue consisted of cells sprouting a variety of tangled projections that formed a continuous network known as the *reticulum*. According to this view, unlike other organs of the body, which could be separated into distinct components, the brain and nervous system could not be disentangled into fundamental and distinct units. Camillo Golgi analyzed nervous tissue following his own staining techniques and noticed that nervous tissue cells had two different kinds of projections: a cluster of short fibers that sprang and branched in many directions and a long cable that didn't branch very much. He noticed that the bodies of the individual cells, although branching near other similar cells, did not in fact fuse to form a continuous reticulum, but so accepting was he of Gerlach's description that he convinced himself that the long connections sprouting from the cells probably still formed a continuous path at some point he could not yet see.

It was Cajal who first realized the individuality of the cells in the nervous system and, more importantly, the implications of what such structural organization might mean. The reticulum description of the nervous system was a monolithic representation impervious to external prodding, threatening to forever keep its operational secrets hidden in a singular mess of tissue. The idea that the nervous system is instead

---

2. Cajal's autobiography, *Recollections of My Life* (*Recuerdos de mi vida*), trans. E. H. Craigie with J. Cano (Cambridge, MA: MIT Press, 1989), quoted in Marina Bentivoglio, "Life and Discoveries of Santiago Ramón y Cajal," NobelPrize.org, Apr. 20, 1998, https://www.nobelprize.org/prizes/medicine/1906/cajal/article/.

made up of billions of individual cells helps, in many ways, break the problem of understanding how it could function into fundamental building blocks and forms the basic principles of today's understanding of the nervous system's organization.

Between 1894 and 1904, Cajal developed one of his most important works, *Textura del Systema Nervioso del Hombre y los Vertebrados* (*Texture of the Nervous System of Man and the Vertebrates*). This work contained detailed analyses and illustrations, made possible by Cajal's artistic affinities and preparation in his younger years, of nerve cell organization and nerve cell structures. His illustrations continue to be reproduced in neuroscience textbooks even to this day. One of Cajal's most important contributions is known as the *law of dynamic polarization*. This law states that nerve cells are polarized by splitting their main function between two distinct parts of their bodies, the input and the output parts. Cells receive input signals on their bodies and dendrites; they generate output signals through the axons.

Framing the humble neuron as a discrete system capable of receiving inputs and producing an output formed the basic principle of how neural information is propagated in the brain. For their contributions to neuroscience, Golgi and Cajal shared the 1906 Nobel Prize in Physiology or Medicine. Golgi's Nobel lecture included a description of his view that neurons form reticular networks. Interestingly, this assertion was then contradicted entirely by Cajal's Nobel lecture, which necessarily focused on the role of the neuron as a distinct unit of the nervous system, far from the concept of a singular, inaccessible reticulum. Cajal continued to fight for his discoveries and disseminate his ideas until his death in 1934. Those ideas form the foundations of the modern understanding of neural information processing, and his contributions are central to artificial neural networks.

Figure 1.2 Tumor cells of the covering membranes of the brain, 1890. *Cajal Institute (CSIS), Madrid.*

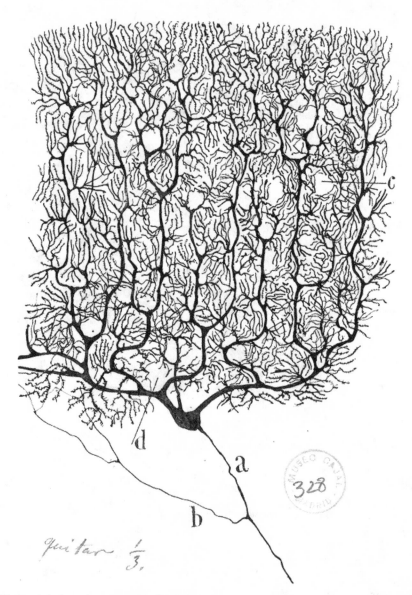

Figure 1.3 A purkinje neuron from the human cerebellum, ca. 1900. *Cajal Institute (CSIS), Madrid.*

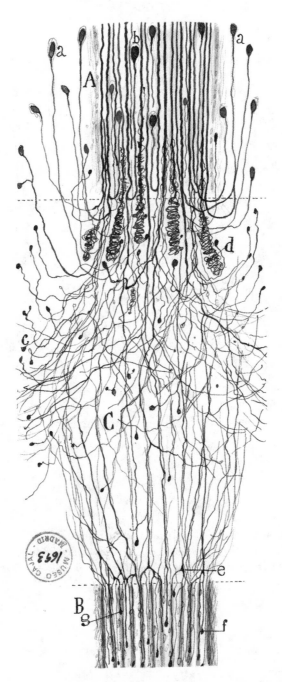

Figure 1.4 A cut nerve outside the spinal cord, 1913. *Cajal Institute (CSIS), Madrid.*

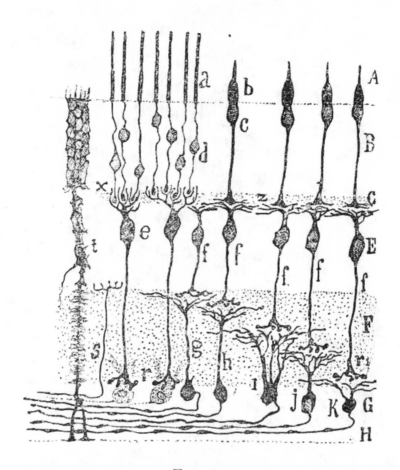

FIG. 27.

Coupe transversale de la rétine d'un mammifère.

Figure 1.5 "Coupe transversal de la rétine d'un mammifère" (Cross section of the retina of a mammal). *Illustration from* Les nouvelles idées sur la structure du système nerveux: chez l'homme et chez les vertébrés *(Paris: C. Reinwald, 1894), 112.*

It's difficult to imagine the full impact of the functions of such a simple unit, which takes an input signal and produces an output signal. It's hard to conceive that when aggregated over a network of billions of such connections we might get intelligence or even conscious behavior. Scientists aiming to explain how activities in the brain relate to overt behavior in the individual have puzzled over this problem for centuries, and they continue to do so today. To understand how complex systems work, it is important to prod them, measure them, and alter their organization in ways that allow us to ascertain a correlation between output behavior and the state of the system at each point in time. Clearly, we are limited in our ability to perform these kinds of investigative and analytical work when it comes to biological (especially human) brains, as changes in these areas typically come with severe adverse effects in the individual. The need to understand biological neural information processes birthed a new set of tools that eventually heralded a new age of artificial intelligence. But it all started with the need to understand the brain. The holy grail was to build a model of the brain that we could tinker with.

GATEWAYS AND GATES

In 1943, Warren McCulloch and Walter Pitts published a research paper titled "A Logical Calculus of Ideas Immanent in Nervous Activity," where the authors proposed the first model of an artificial neuron. It was a very simple system: a node with inputs that could be either 0 or 1 and a firing threshold, which could be any value and served as a mechanism for determining when the neuron would fire. To fire, the aggregate of the input values into the neuron needed to match or exceed the threshold.

For example, consider a neuron with two inputs X_1 and X_2, a threshold value of 2, and an output Y. If both inputs X_1 and X_2 are 1, then 1 + 1 = 2, which matches the threshold value of 2; therefore, the neuron fires, and Y = 1 (fig. 1.6). If both inputs are 0, then 0 + 0 = 0, which does not match or exceed the threshold value of 2; thus, the neuron does not fire, and Y = 0. Moreover, if either input signal is 0, then 1 + 0 = 1, which does not meet the firing threshold of 2. Again, the neuron would not output a signal, and the output value Y would be 0. This was the first functioning artificial neuron, and whenever you see your little Roomba headbutting a table leg in apparent artificial frustration, take a moment to think back to Santiago Ramón y Cajal and the McCulloch-Pitts neuron; the little Roomba would not exist without them.

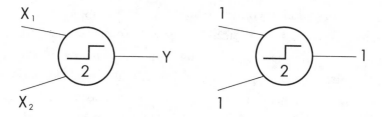

Figure 1.6 A McCulloch-Pitts neuron with two inputs, a firing threshold of 2, and one output. In the example on the right, both input signals have a value of 1.

The McCulloch-Pitts neuron is a simple system meant to model the most elemental functions of biological neurons. In 1943, thanks to previous work by Cajal and Sir Charles Scott Sherrington (a 1932 Nobel laureate and author of *The Integrative Action of the Nervous System*), we knew that biological neurons received input signals via the dendrites and that based on some internal threshold or a more nuanced calculation, the neuron would either send out a signal through the axon or not. If you notice, the McCulloch-Pitts neuron approximates the basic elements in this process. It expects a set of values through the input connections, and based on a simple internal state (a threshold value), it outputs either a 1 or a 0. The question is how useful a system like this is. Can we really do anything with something so simple? It turns out that we can.

One of the more useful things we can do is implement a special set of logic functions known as *logic gates*. Interestingly, all modern advanced electronics and computing are based on logic gates. They help us transform 0s and 1s from signals traveling in a wire to states upon which we can make decisions. To build the intuition for how simple systems like neurons can perform complex tasks when integrated in complex networks, it will be helpful to understand logic gates. The simplest logic gates are NOT, AND, and OR.

AND Gates

The AND logic gate has two input wires (fig. 1.7). When both input wires are "on" (i.e., both wires have a signal representing the value 1), the output of the AND gate is 1. If both input wires are "off" (i.e., both wires have no signal, representing the value 0) or even if either one of the input wires is "off," the output of the gate is 0. For the AND gate to output a signal of 1, both input 1 *and* input 2 must be 1 (table 1.1).

How can we utilize this capability in a more practical use case? Suppose you are building a security feature common in industrial machines—like a hydraulic press—where to ensure that the operator's hands are off the danger area while the machine is operating, there are two buttons that must be constantly pressed by the operator for the machine to work. In this system, you might add a logic AND gate with input wires connected to the two buttons in the control panel and connect the output of the logic gate to the machine's circuit controller. Only when both buttons are pressed will both inputs to the AND gate have a value of 1, and only then will the AND gate output a 1, signaling that the heavy machinery can proceed to operate. If neither of the buttons is pressed or if only one button is pressed, the AND gate will output a value of 0, and the machine will remain off.

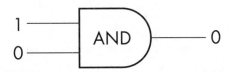

Figure 1.7 An AND logic gate with two inputs and one output. If either of the inputs is 0, the output is 0. If both inputs are 1, the output is 1.

Table 1.1 AND Gate and Possible Input/Output Combinations		
Input 1	Input 2	Output
0	0	0
0	1	0
1	0	0
1	1	1

NOT Gates

The NOT logic gate is probably the simplest of the logic gates, and it functions as a signal inverter (fig. 1.8). It has one input wire and one output wire. If the input is 0, the gate outputs a 1. If the input is 1, the gate outputs a 0 (table 1.2).

We can apply this behavior to a wide array of use cases. For example, suppose we want to design a sensor that can tell us if the fuel inside the fuel tank of a car drops below a certain level. We might set up a fuel sensor connected to a NOT gate. While the sensor is submerged in the fuel, the input to the NOT gate is 1, and the output is 0. An output signal of 0 is interpreted by the control circuit as the fuel level being fine. As soon as the fuel level drops below the sensor's reach, the input signal to the NOT gate becomes 0, and the output turns to 1. This causes the NOT gate to send a signal to the control board indicating that we are low on fuel.

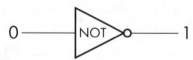

Figure 1.8 A NOT logic gate, with one input and one output. If the input is 0, the output is 1. If the input is 1, the output is 0.

Table 1.2 NOT Gate and Possible Input/Output Combinations	
Input 1	Output
0	1
1	0

OR Gates

The OR logic gate is similar to the AND gate (fig. 1.9). It has two input wires and one output wire. It outputs a signal when either input wire is on (table 1.3).

Again, we can take advantage of this function by acting on an event based on a sequence of past events. Say that we are tasked with implementing a payment system for a public transportation service, a subway, and there are two methods customers can use to pay the fare. They can choose to scan a payment card and transfer the funds electronically, or they can choose to pay in cash. The payment system can be connected to an OR gate where the electronic payment system and the coin collector system are connected to the input wires of the logic gate, and the output of the OR logic gate is connected to the controller board for the turnstile. While the customer has not paid either electronically or using cash, the OR gate continues to not output a signal (output is 0), and the turnstile remains locked. As soon as the customer pays either electronically *or* using cash, a signal (output is 1) is sent to the turnstile's controller board that lets the customer pass through.

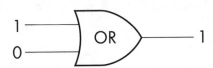

Figure 1.9 An OR logic gate with two inputs and one output. If either input is 1, the output is 1. If both inputs are 0, the output is 0.

Table 1.3 OR Gate and Possible Input/Output Combinations		
Input 1	Input 2	Output
0	0	0
0	1	1
1	0	1
1	1	1

Logic Gate Systems

Hopefully you can begin to see how powerful individual logic gates can be. And just as individual neurons are combined to form powerful networks, we can also combine logic gates by using the output of one gate as the input to another gate.

Consider an automated process for receiving travelers at an airport in a country belonging to the European Union. (Clearly this is an unrealistic example, but bear with me.) There are two ways to get into the country: if you are a citizen of the EU, you can pass right in; if you are not a citizen of the EU, you must show a valid visa and international passport. We could build a system that lets travelers through based on an AND gate combined with an OR gate (fig. 1.10). The AND gate is connected to one of the OR gate's inputs. The AND gate outputs a 1 if you have an international passport *and* a valid EU visa. The other input of the OR gate is set to 1 if you are a citizen of an EU nation. Following this logic, if you are a citizen of the EU, at least one of the inputs into the OR gate is set to 1; therefore, the output of the OR gate is 1, and you are allowed through. If you are not a citizen of the EU, one of the inputs into the OR gate is 0, so for the OR gate to output 1, the other input must be 1. The other input depends on the output of the AND gate, which is only 1 if you have both an international passport and a valid EU visa.

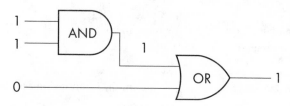

Figure 1.10 An AND gate and an OR gate automatically controlling access into the EU.

We can continue to combine logic gates and build complex circuit systems. In fact, if we combine enough logic gates (millions of them), we can build modern computers. Logic gates are the building blocks of the control flow and decision systems that form the backbone of programming languages. Software consists of a series of instructions that a computer can understand. These instructions are sometimes predicated on certain logical conditions (e.g., "if this, then that"). Those conditions are made possible by logic gates. When we read about new computer processors—CPUs or GPUs—sporting millions of transistors, this is essentially what those numbers mean. Transistors are used to a large extent to implement logic gates, so millions of transistors equate to millions of logic gates. So why, again, are we talking about logic gates?

It turns out that we can build logic gates using McCulloch-Pitts neurons, and logic gates were already well-known constructs in the 1950s.[3] This provided scientists with the realization that if you could build logic gates with artificial neurons, then these neurons could be used to build other kinds of powerful systems, maybe even a model of the brain. It gave scientists permission to continue investigating artificial neurons.

Fine-Tuning the Neuron

A limitation of the McCulloch-Pitts neuron is that it takes binary inputs and outputs binary values. Even in the 1940s, we knew that inputs into biological neurons were not always binary values (i.e., full-intensity signal or zero-intensity signal). In biological neurons, the signals are nuanced and cover a range of intensities. Furthermore, the decision on whether the neuron should fire, also referred to as the *activation function* of the neuron, used in the McCulloch-Pitts neuron is the threshold function, which again we knew doesn't resemble biological neurons very closely (fig. 1.11).

In 1949, Donald Hebb, a Canadian psychologist researching how neurons contribute to the learning process, wrote a book titled *The Organization of Behaviour*. In this book, Hebb introduced the

---

3. Charles Babbage (widely regarded as the father of the digital computer) used early (mechanical!) logic gates in the 1830s in his "analytical engine." In the 1920s and 1930s, more modern logic gates were invented for the computers of the time.

theory of *Hebbian learning*, which revolutionized the way neural information processing was understood and helped change the way that artificial neurons were implemented. In his book, Hebb says, "When an axon of cell A is near enough to excite a cell B and repeatedly or persistently takes part in firing it, some growth process or metabolic change takes place in one or both cells such that A's efficiency, as one of the cells firing B, is increased."[4] In this statement Hebb proposes that as neurons fire together, not only is the connection between the neurons strengthened, but this very operation is a fundamental step in the learning process. This suggests that weights should accompany the connections to neurons, where the weights modulate the importance of each connection over time. That is, the information flowing through a neural network is not dictated by the input values alone; the connections between the neurons themselves are strengthened or weakened through the learning process by adding a weight value to each connection. The weights affect how much of the input value into that connection is received by the neuron. The addition of weights means that neural networks are not obdurate, unchanging things. Instead, they can be adjusted, and information can be routed and rerouted depending on how the connections between the neurons are strengthened or weakened, much like how water might flow downriver depending on which obstacles it finds.

Take a moment to ponder this discovery. It was extremely important. The addition of weights to the connections was akin to adding a frequency tuning dial to a radio. A radio that is built tuned to a specific frequency, with no dial, is like the McCulloch-Pitts neuron. We must rebuild the neurons to change their outputs, just as we must rebuild the radio to change the frequency it's tuned to. The addition of weights to the neuron suggests that we can move the tuning dial on the neurons and change their behavior at run time, much as we can change the frequency on a radio by turning the dial.

---

4. D. O. Hebb, *The Organization of Behaviour: A Neuropsychological Theory* (New York: John Wiley & Sons; London: Chapman & Hall, 1949), 62.

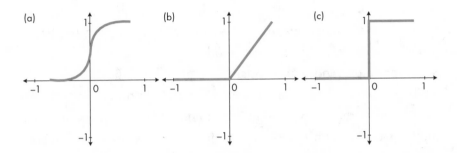

Figure 1.11 Three activation functions we refer to often in this chapter. The sigmoid activation function (*a*) outputs a value between 0 and 1 for all its inputs, with an input 0 producing 0.5 as output. The rectified linear unit activation function (*b*) outputs 0 for all negative inputs, and outputs the unmodified input value for all positive inputs. The threshold function (*c*) outputs a 0 for all input values below the threshold value; for input values equal to or greater than the threshold, it outputs a 1.

THE PERCEPTRON

Frank Rosenblatt, an American psychologist working in 1958 at the Cornell Aeronautical Laboratory as part of a project funded by the U.S. Office of Naval Research, studied the work on the artificial neuron done by McCulloch and Pitts and recent discoveries by Hebb and developed the *perceptron*. The perceptron was a modification of the McCulloch-Pitts neuron that incorporated Hebb's weighted input connections, as well as allowing each individual input to take a value *between* 0 and 1, instead of the binary (all-in/all-out) approach of the McCulloch-Pitts neuron. These changes more closely aligned the perceptron to biological neurons.

The history of artificial neural networks is riddled with subtle but crucial discoveries. The addition of weights and real numbers (numbers with fractions that can measure continuous quantities; e.g., all numbers between 0 and 1, not just 0 or 1) hinted that it might be possible to build general models of the brain that could be adjusted to solve general sets of problems, as opposed to specific problems. As

Rosenblatt described in his 1961 book, *Principles of Neurodynamics*, there are two approaches to creating models for studying the brain: monotypic models and genotypic models.

Monotypic models are similar to our pretuned radios with no dials, while genotypic models are adaptable and allow tuning to different frequencies without having to build a different radio. In the monotypic model, we start with functional requirements of some "psychological function,"[5] for example, a well-defined recognition algorithm with specific input/output conditions. That is, in the monotypic model, we find a psychological function that we want to model. Say we want to model how smell is processed in the brain. We might design an artificial system capable of detecting certain types of smells and then design a specific reaction to each smell. Once our artificial system is functioning, we monitor the process of detecting smells and reactions to each smell in a human volunteer, and we try to find where the artificial and biological systems are similar. The monotypic approach for studying the brain starts with a specific set of inputs and specific set of desired outputs, builds a system that can map those inputs to the desired outputs, and finds similar behavior in biological systems. We might then update our model based on what we see in the biological systems. This approach better aligns with the McCulloch-Pitts neuron. It is a purposeful approach where we know exactly the kind of solution we are building.

In the genotypic approach, instead of starting with well-defined functions and comparing our artificial system to the brain, we start with a set of generic learning rules and build a more general algorithm that can learn to model any set of problems following the same training procedure (rather than a training procedure or design specific

---

5. Frank Rosenblatt, *Principles of Neurodynamics* (Buffalo, NY: Cornell Aeronautical Laboratory, 1961), 11. All other quotes in this section are drawn from Melanie Lefkowitz, "Professor's Perceptron Paved the Way for AI—60 Years Too Soon," *Cornell Chronicle*, Sep. 25, 2019, https://news.cornell.edu/stories/2019/09/professors-perceptron-paved-way-ai-60-years-too-soon.

to the problem at hand). For example, instead of building a system for detecting a defined range of smells, another system for detecting a defined set of tastes, and yet another system for detecting a desired range of sounds, the genotypic approach aims to find a generic design that can detect input stimuli that could be sounds, tastes, or smells. The genotypic approach lends more plasticity to the design of our artificial networks because the design does not have to closely implement a set of initial requirements; instead, the network can learn to adjust itself to develop the requirements that help it meet its output goals. By incorporating both real-number values into the network connections (numbers between 0 and 1), as opposed to the Boolean (0 or 1, true or false) nature of the McCulloch-Pitts neuron, and the weighting of synaptic connections, the Rosenblatt perceptron is better equipped for implementing genotypic models.

Rosenblatt viewed decision-making and intelligence in the brain as following a set of statistical and probabilistic algorithms where, instead of mapping a set of input stimuli to a specific set of output psychological behaviors (monotypic models), the brain maps classes of inputs to classes of outputs (genotypic models). There is a very important distinction between these two approaches. If the brain functioned exclusively based on mapping specific inputs to specific outputs—the collection of predefined algorithms necessary to perform all the functions we perform daily and throughout our lives—we would need a staggering collection of discrete algorithms in our brains. This would make the dream of creating artificial systems capable of emulating human-level intelligence almost certainly impossible. If, instead, the brain functioned as a statistical system mapping classes of inputs to classes of outputs, we would not need a specific algorithm for every function we perform; we'd just need an algorithm for each class of functions we perform. This at least offers a reduction in the number of systems we need to emulate if we want to artificially build human-level intelligence.

Of course, the brain (and nature) need not listen to Rosenblatt, and in fact, it may be that the brain employs both approaches, some specific algorithms for special functions and generic algorithms for most other functions. Rosenblatt's point was that if we want to have a shot at building a system that resembles human intelligence, and since we don't really know how the brain exactly works anyway, let's assume that it works in the way that offers us a greater chance at emulating it and see how far we get. It is important to note that Rosenblatt, at least initially, did not build the perceptron with the purpose of creating artificial intelligence for its own sake. His primary goal was to build a system that he could tinker with and alter to help him better understand our brain.

So how did this perceptron really work? The Rosenblatt perceptron functioned as follows: For a perceptron with five inputs X_1, X_2, X_3, X_4, X_5, each input connection (analogous to the biological neurons' synapses) has a weight $(w_1, w_2, \ldots, w_5)$ associated with it. The weights are also real numbers, which means they can take any value between 0 and 1 (note that the values can also be greater than 1, but they are typically normalized to the 0–1 range). The process for calculating the output of a perceptron is to perform a weighted sum over the inputs and apply the activation function on the result. The classic activation function for the perceptron was the threshold function, just as it was with the McCulloch-Pitts neuron, but, as we saw in figure 1.11, there are many activation functions that could be used. In the case of the threshold function, we check whether the result of the weighted sum is greater than the threshold value. If it is, the neuron outputs a 1; otherwise, it outputs a 0. A *weighted sum operation* is performed as follows:

$$V = \left(\sum_{j=1}^{n} w_j x_j \right) + b$$

The formula might look complicated but remember that Σ simply means a sum, so the equation is saying that the output value (V) is equal to the sum of the input values (x) times the weight (w) of the connections, for all inputs (j), plus b. We are going to largely ignore b until chapter 3. For now, we just need to know that it's called a *bias*, and as we can see, it doesn't contribute a lot to the output compared to the input connections and the weights.[6] For the next two chapters, we want to build an understanding of how neural networks work at the neuron level, so we are going to concentrate on the most important part of the input/output process—that is, the inputs, the weights, and the activation functions. Once we calculate the weighted sum over the inputs, we take the resulting value V and perform the activation function calculation $O = f(V)$, which gives us the output value (O) for the neuron.

To calculate the output of the perceptron shown in figure 1.12, we perform the following operation, $O = f(w_1x_1 + w_2x_2 + w_3x_3 + w_4x_4 + w_5x_5) + b$, where f is the threshold activation function with a threshold value of 2. The neuron outputs a 1 if the result of the summation is greater than or equal to 2; otherwise, it outputs a 0. This all sounds great, but how do we get these magical weight values? The weights of the perceptron are initially chosen randomly; they then get adjusted during the training phase of the model when the output is evaluated against expected results, and updates are made to the weights to ensure that the model performs better during each iteration of the training cycle. But more on this later.

---

6. The bias is certainly a necessary term, and depending on the scale of the data, it can become quite important. We are temporarily ignoring this term, however, because the more involved calculations concern the connection weights. The math is already complex enough, and accounting for the bias in our calculations might confuse novice readers. But note that, as described by the equation, we would add the bias term after every set of per-layer transformations.

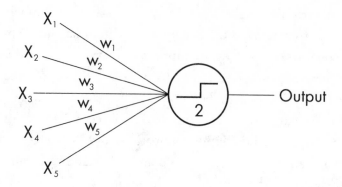

Figure 1.12 A Rosenblatt perceptron with five input connections. Each connection has a weight (*w*) associated with it. The value of the weight can be any number between 0 and 1. The output of the perceptron depends on the weighted sum operation passing the threshold test.

If you are wondering whether this perceptron has practical applications or if it's just a scientific novelty item, it is definitely useful in solving certain types of problems. The perceptron was first implemented in software for the IBM 704. It was a computer the size of a room and used punch cards as a user interface. That perceptron was trained to distinguish between punch cards marked on the left and punch cards marked on the right—an incredible achievement for the time. Rosenblatt called the perceptron "the first machine which is capable of having an original idea." The first implementation of the perceptron as an automation tool was the Mark 1 perceptron, developed in 1958; in this case, it was a machine, as opposed to software. It was designed for image recognition capable of analyzing input images of 20 pixels width by 20 pixels height. The input into the perceptron was a camera with 20 × 20 calcium sulfide photocells. The 400 (20 × 20) photocells were randomly connected to the input channels of the perceptron; that is, the system employed a perceptron with 400 inputs. The weights of the connections to the perceptron were encoded by an array of special resistors called potentiometers, which could be actively adjusted to vary the resistance and in turn vary the voltage.

In computer systems, values of 0, 1, and anything in between are represented using voltages. The potentiometers were useful to allow for adjusting each connection weight at run time and representing any value between 0 and 1. The potentiometers were adjusted during the learning phase using electric motors to dial into the right voltage value for each weight. In 1959, Bernard Widrow and Marcian Hoff of Stanford University applied advances in the perceptron to the first implementation of neural networks to solve a real-world problem. These were artificial neural networks trained to eliminate noise from phone lines. Named ADALINE and MADALINE (which stand for ADAptive LINear Elements and Multiple ADAptive LINear Elements, respectively), these systems are still in use today!

It seemed that all was going well with artificial neural networks, and real artificial intelligence—complete with humanlike androids ready to do our bidding—was just over the horizon. So what happened?

Right around this time, enthusiasm over a possible new era of human-machine interactions seemed to reach a boiling point, with the *New York Times* running an article titled "NEW NAVY DEVICE LEARNS BY DOING: Psychologist Shows Embryo of Computer Designed to Read and Grow Wiser" and the *New Yorker* writing, "Indeed, it strikes us as the first serious rival to the human brain ever devised." Rosenblatt himself didn't do much to ground such overhyped expectations with statements like, "We are about to witness the birth of such a machine—a machine capable of perceiving, recognizing, and identifying its surroundings without any human training or control."[7]

CALL OF DUTY RESCUES THE NEURAL NETWORK

Unfortunately for Rosenblatt, and artificial neural networks in general, the excessive attention these systems garnered in a short period served to irk artificial intelligence researchers pursuing more traditional

7. Lefkowitz, "Professor's Perceptron."

approaches. Most notable among these researchers was Marvin Minsky, who in 1969 coauthored a book titled *Perceptrons* (with Seymour Papert). In the book, Minsky took aim at the perceptron and its inability to distinguish between classes that are not linearly separable. The perceptron that Rosenblatt implemented for image recognition using a 20×20 photocell camera consisted of a single-layer perceptron, and single-layer perceptrons are limited to linearly separable problems. That is, if you think of a classifier as a line that separates two classes of objects (say, blue dots from red dots), a line can only separate these two classes of objects if they are arranged in such a way that a line can bisect the space in which they live. Take a group of blue marbles and red marbles spread over a tabletop: we could position a stick between the red marbles and the blue marbles only if the marbles are already grouped by color. If the marbles are mixed on the surface of the table, a stick could not be placed in a way that separates blue marbles from red marbles. This is the attack that Minsky launched on the single-layer perceptron. More specifically, he cited that the single-layer perceptron could not implement the EXCLUSIVE OR (XOR) logic gate.

We discussed logic gates earlier in the chapter. If we refer to the description of the OR gate, we saw that the gate outputs a 1 if either input X_1 or X_2 is 1, including the case where both inputs are 1. The XOR gate outputs a signal of 1 only if either input X_1 or X_2 is 1, not if both are 1; this is the "exclusive" part. It turns out that the single-layer perceptron cannot implement this logic gate. To understand why, it helps to think of the perceptron as a linear classifier. If we graph out the possible gate outputs depending on the possible gate inputs, we see that for the AND and OR gates, we can easily draw a line that separates the 0 and 1 outputs, but for the XOR gate, we cannot draw a single line that separates the two classes of outputs (fig. 1.13). This has profound implications because the kinds of problems we need to solve in the real world are most often not linearly separable.

To be fair, Minsky was aware that a plausible solution to this problem was to stack more perceptron layers together and create a more complex neural network. The problem, as Minsky pointed out, was that a neural network complex enough to solve nonlinearly separable problems like the classification of images of higher resolution would take an impractically long time to train. That is, it would take too long to calculate the correct weight values for each connection between the neurons in each layer. This criticism ushered in an era that is known as the "AI Winter." For the next twelve years, government funding for neural network research all but dried up. This is the problem with hype, and science is not immune to it. Scientific progress typically happens along a smooth curve, with new findings slowly building up from previous discoveries; that is, it's a gradual process. Only seldomly does scientific progress experience a stepped advance where a new discovery significantly improves on established wisdom. When some scientists and the media invariably overhype new technologies or discoveries, the hype contributes to the technology's demise when predictably the result does not match the public's expectations. It's easy to play historical revisionism, but perhaps Minsky's attack would not have been so lethal if instead of expecting the perceptron to be the end-all-be-all solution to AI, we had just seen it as a gradual step in the right direction.

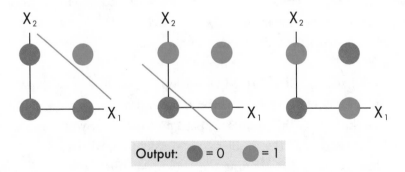

Figure 1.13 Possible outputs of the AND, OR, and XOR logic gates. The AND gate outputs a 1 only when both X_1 and X_2 are 1. This is shown as the *red circle*; otherwise, it outputs a 0 (*blue circle*). The outputs of an AND gate are linearly separable: we can draw a single line that separates the two classes of outputs, 0 and 1. The same is true for the or gate. But for the XOR gate, we cannot draw a single straight line to separate the classes of outputs: the outputs of the XOR gate are not linearly separable.

Thankfully, neural networks would make a resurgence. The mood started to change again in 1982 when John Hopfield presented the Hopfield Net at the National Academy of Sciences. Also, at around this time, Japan announced a fifth-generation computer project aimed at bolstering national research in artificial intelligence, including neural networks. This was announced at the U.S.-Japan Joint Conference on Cooperative/Competitive Neural Networks, reinvigorating the U.S. Defense Department's interest in neural network research. Funding soon started to flow again in an effort to maintain the United States' technological edge in the face of Japanese advancements. Periods of inflated expectations and funding followed by disillusions and dried-up funds would continue into the 1990s and early 2000s. In 2012, Alex Krizhevsky, Ilya Sutskever, and Geoffrey E. Hinton, of the University of Toronto, authored a paper titled "ImageNet Classification with Deep Convolutional Neural Networks" that helped reignite interest in AI, which has maintained to the present, and this time it promises to endure a bit longer. Neural networks are already solving real-world problems, and although what their future looks like and what they might still be capable of are subject to much speculation, the solutions they are providing will continue to serve as fertile ground for sustained development. The hype situation has not improved much over the past few decades, but something unexpected has facilitated neural networks' persistence, in full research fury, almost a decade after being rediscovered.

So what changed in 2012? In the 1960s, researchers knew that if you stacked a series of single-layer perceptrons and created a more complex network known as the multilayer perceptron (MLP), you could solve nonlinear problems. The problem, as Minsky pointed out at the time, was that the process of training the network—in other words, finding the correct value for each connection weight—would take hundreds or thousands of years of computational time, which made it impractical. Computers were much more limited, however, in the 1960s. Also, researchers were thinking of computations as happening within CPUs. CPUs are the main processing component of a computer system, and we can think of instructions executing in CPUs as executing sequentially. That is, if we want to perform ten operations, the CPU starts with the first operation, proceeds to the second operation, and so on. When the problem we are trying to solve consists of performing millions of computations (as is the case with complex neural networks), if these computations are done sequentially, the time required to train the network will make it impractical even for the fastest modern CPUs. Still, as CPUs became more powerful in the late 1990s and early 2000s, some large CPU clusters known as *supercomputers* were able to crunch vast numbers of computations, and research into neural networks slowly picked up again. Considering that not everyone has access to a supercomputer, however, neural networks remained at the fringe of artificial intelligence research. Luckily though, advancements in a completely different industry were about to pull the humble neural network out of the periphery and place it right into the mainstream.

The gaming industry had been pushing graphics processing unit (GPU) companies to produce faster and more efficient GPUs since the first pixelated computer games appeared in the 1980s. One interesting thing about computer graphics is that operations happen at the pixel level. If we think of a computer game from a pure graphics perspective, what we are looking at is a series of frames that must be drawn on a screen. (Like a still from a movie, a frame is the image that fills your screen and shows the evolving scene.) The frames are made of thousands (or millions) of pixels. Increasing performance and efficiency in computer graphics means creating processors that can parallelize the work that's necessary to draw those pixels. This meant that in 2012 the most powerful processors people could access outside of supercomputers weren't CPUs; they were GPUs. For the work we need to perform to train a neural network, we can think of CPUs and GPUs as powerful calculators. The difference between the CPU and the GPU is that the CPU performs all its calculations sequentially, and the GPU can perform thousands of operations in parallel; therefore, a GPU can process a set of operations much more quickly than a CPU.

Krizhevsky, Sutskever, and Hinton were aware of this when they set out to train a complex neural network using a GPU. Their neural network implementation, named AlexNet, competed in the ImageNet Large Scale Visual Recognition Challenge in 2012 and achieved state-of-the-art results when compared with any image recognition model that year. It is important to note that, along with advancements in GPUs, ImageNet itself was another equally important benefit that Krizhevsky, Sutskever, and Hinton enjoyed in 2012, which did not exist decades earlier. In 2006, Fei-Fei Li, at that time a researcher at Princeton University, began working on an interesting project to build a vast data set of natural images along with annotated labels. The label of an image describes the subject of that image and can be used to teach AI models to correctly classify images according to their labels. Thus, this data set presents an invaluable tool for training AI models in visual recognition tasks. Today, ImageNet has over 14 million annotated images, and it is a keystone in the development of many computer-vision applications.

Interestingly, the architecture of AlexNet was similar to a convolutional neural network (CNN) proposed by Yann LeCun in 1989. Unfortunately, in 1989 LeCun didn't have access to ImageNet or a GPU powerful enough to train a deep (many-layered) model. What GPUs did, though, was democratize access to powerful calculators that could perform thousands of operations in parallel. Practically anyone can afford a GPU and set up a system to train a neural network. Suddenly, progress could be made in every lab, without running into bottlenecks when trying to access specialized supercomputers. This, along with ImageNet and similar data sets created by researchers with the help of the internet, allowed for an explosion in neural network research: all sorts of existing architectures were explored, and new architectures also emerged. AlexNet is a CNN, and we look at CNNs in the next chapter when we discuss computer vision in detail. Now we examine the model that started it all: the multilayer perceptron.

PEELING BACK LAYERS

We have seen how single neurons work. We learned about their history and how different discoveries helped transform the design of artificial neurons. Now we want to see how neural networks function. We want to design a network that's made up of several layers of artificial neurons, and we want to understand what happens at each neuron.

Let's consider a simple MLP, also known as a *fully connected* neural network. We can create a simple model with three input nodes, followed by a four-neuron layer, followed by a two-neuron layer, and ending with a single output neuron (fig. 1.14). The layers between the input and the output—that is, the meat of the neural network— are called the *hidden* layers. As noted in the introduction, the term *deep learning*, or *deep neural networks*, refers to neural network models that have more than one hidden layer. In our case, the model we are discussing has two hidden layers.

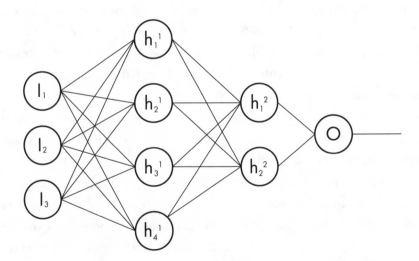

Figure 1.14 A multilayer perceptron (MLP) with two hidden layers. Three input nodes are connected to the first hidden layer (h^1), and h^1 is connected to the second hidden layer (h^2), which is connected to the output neuron.

The first hidden layer, the one with four neurons, we are going to call h^1, and the second hidden layer we are going to call h^2. MLPs are also known as fully connected neural networks because each neuron of every layer is connected to every neuron of the next layer. Input node I_1 is connected to $h_1^1, h_2^1, h_3^1, h_4^1$. Input node I_2 is also connected to $h_1^1, h_2^1, h_3^1, h_4^1$, and so is input node I_3. Then, the output of h_1^1 is connected to h_1^2 and h_2^2. The output of h_2^1 is connected to h_1^2 and h_2^2, and so are the outputs of h_3^1 and h_4^1. The output of h_1^2 is connected to the output neuron O, as is the output of h_2^2 (fig. 1.14). The output of each neuron depends on the values of each connection associated with it and an activation function. We call the values for each neuronal connection *weights*, and the activation function dictates the range of values that the neuron can output (or in some cases, it may limit the output of a neuron until a certain input value range is crossed).

The purpose of an activation function is to introduce nonlinearity into the calculations the neural network is performing. We discuss why nonlinearity is important in the section below on vector spaces. For now, we can just recall that biological neurons exhibit similar nonlinear behavior. Hebb showed that the relationship between the input signals at the dendrites and the output signal at the axon does not follow a linear path, where the output value is simply the sum of the input signals. The output of the neuron is regulated by some internal function that considers the input signals and transforms them by some process. The artificial neurons in our model emulate this process by using mathematical activation functions. Researchers have come up with many activation functions, and depending on the problem we are trying to solve, certain activation functions are better suited than others. Two popular activation functions are the *sigmoid* and the *rectified linear unit* (ReLu) activation functions. The sigmoid function outputs values between 0 and 1, so it's widely used for use cases where we want the neural network to output a probability; probabilities range between 0 and 1, with 1 as the maximum value we can achieve (a 100% chance). The ReLu activation function is one of the most popular activation functions in use today. It simply looks at the input value, and if it's a negative value, it outputs a 0; if it's a positive value, it outputs the input value. For the following example, we use the ReLu activation function for all neurons except the output one. This is a common approach for *binary classification* neural networks (binary classification means separating data samples into two possible classes, e.g., cat vs. dog). For the output neuron, we use the sigmoid activation function to ensure that the output can be interpreted as a probability. Now let's see how information flows through the neural network.

To start, we calculate the output of the first neuron in h^1. We are going to presume that the neural network has already been trained, and we have the weight values for each connection (fig. 1.15). Let's assume the input vector we want to process is I = [0.2, 0.01, 0.4].

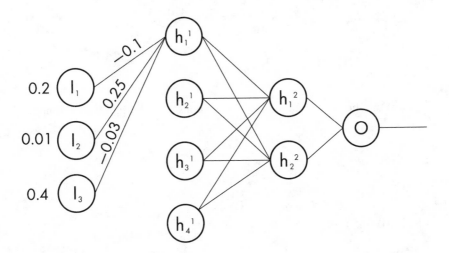

Figure 1.15 The weight values of the connections to h_1^1. Note that we have removed the connections to the other neurons in h^1 only for visibility reasons.

We then execute a weighted sum operation between the input vector and the connection weights:

$h_1^1 = ReLu(w_{1,1}^1 I_1 + w_{1,2}^1 I_2 + w_{1,3}^1 I_3)$

$= ReLu(-0.1*0.2 + 0.25*0.01 + (-0.03*0.4))$

Recall that ReLu functions output 0 for negative values.

$= ReLu(-0.0295)$

$= 0$

We calculate the output of the second neuron in h^1 in the same fashion (fig. 1.16).

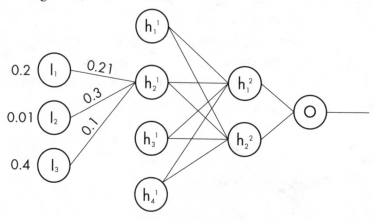

Figure 1.16 The weight values of the connections to h_2^1 (as in fig. 1.15, simplified for visibility).

$h_2^1 = ReLu(w_{2,1}^1 I_1 + w_{2,2}^1 I_2 + w_{2,3}^1 I_3)$

$= ReLu(0.21*0.2 + 0.3*0.01 + 0.1*0.4)$

$= ReLu(0.085)$

$= 0.085$

The output of the third neuron (fig. 1.17):

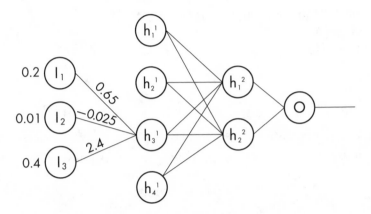

Figure 1.17 The weight values of the connections to $h_3{}^1$.

$h_3{}^1 = \text{ReLu}(w_{3,1}{}^1 I_1 + w_{3,2}{}^1 I_2 + w_{3,3}{}^1 I_3)$

$= \text{ReLu}(0.65*0.2 + (-0.025)*0.01 + 2.4*0.4)$

$= \text{ReLu}(1.1)$

$= 1.1$

The output of the fourth neuron (fig. 1.18):

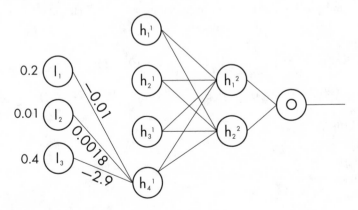

Figure 1.18 The weight values of the connections to $h_4{}^1$.

$h_4^1 = \text{ReLu}(w_{4,1}^1 I_1 + w_{4,2}^1 I_2 + w_{4,3}^1 I_3)$

$= \text{ReLu}(-0.01*0.2 + 0.0018*0.01 + (-2.9)*0.4)$

$= \text{ReLu}(-1.16)$

$= 0$

Now that we have the output of each neuron from h^1, these values become the inputs into the second layer, h^2 (fig. 1.19). To calculate the output of the first neuron of h^2, we proceed as follows:

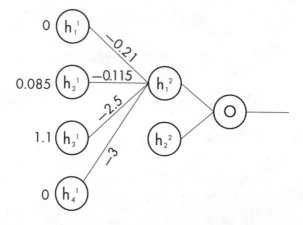

Figure 1.19 The inputs and weight values of the connections to h_1^2. Note that each input into layer h^2 is the output of h^1 [0, 0.085, 1.1, 0], which was calculated in the previous steps.

$h_1^2 = \text{ReLu}(w_{1,1}^2 h_1^1 + w_{1,2}^2 h_2^1 + w_{1,3}^2 h_3^1 + w_{1,4}^2 h_4^1)$

$= \text{ReLu}(-0.21*0 + (-0.115)*0.085 + (-2.5)*1.1 + (-3)*0)$

$= \text{ReLu}(-2.75)$

$= 0$

The output of the second neuron in h^2 (fig. 1.20):

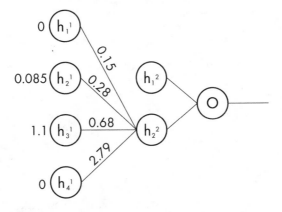

Figure 1.20 The inputs and weight values of the connections to h_2^2.

$$h_2^2 = \text{ReLu}(w_{2,1}^2 h_1^1 + w_{2,2}^2 h_2^1 + w_{2,3}^2 h_3^1 + w_{2,4}^2 h_4^1)$$

$$= \text{ReLu}(0.15*0 + 0.28*0.085 + 0.68*1.1 + 2.79*0)$$

$$= \text{ReLu}(0.77)$$

$$= 0.77$$

Finally, the output of the neural network is calculated using the outputs of h^2 as inputs to the output neuron (fig. 1.21).

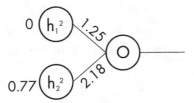

Figure 1.21 The output of h^2 as the input into the output layer, along with the connection weights to the output layer.

Note that for the output layer, we use the sigmoid activation function in this example. The sigmoid function takes the following form: $\frac{1}{1+e^{-(x)}}$

$O = \text{sigmoid}(w_{1,1}^{3} h_{1}^{2} + w_{1,2}^{3} h_{2}^{2})$

$= \text{sigmoid}(1.25*0 + 2.18*0.77)$

$= \text{sigmoid}(1.68)$

$= \dfrac{1}{1+e^{-(1.68)}}$

$= 0.84$

That's it—we did it! We started with an input and a set of predetermined weights, and we processed the input using the neural network and got a result. The sigmoid activation function looks a bit complicated, but we don't have to dwell on it too much. We just need to know that it takes a value as an input and produces an output that's between 0 and 1, which can be interpreted as a probability. At this stage in our progress, we still don't know how to interpret the output value, but that's OK. The purpose of this exercise was to show the operations taking place inside a neural network. Although we are not yet sure what 0.84 means or how it can be useful to us, we have now seen the operations that produce an output, and we can see that they are simple operations. There is no magic happening in the individual calculations themselves; for the most part, we are just performing multiplications and additions. These multiplications and additions between the inputs and the weights of the neuron connections are what we call a weighted sum operation, and these operations are the fuel that powers the predictions in artificial neural networks.

In common parlance, we call the output of a neural network a *prediction*. This is because, in a general sense, what the neural network is doing is outputting a value that has a certain probability of being correct. Based on previous data that the network saw during training, the neural network builds a certain model of the world. When we present it with a new input, it outputs a prediction. In classification use cases, the prediction is typically a probability that the input belongs to a particular class of objects. Not all the use cases where we employ neural networks today are classification problems, however; some are *regression problems* (a type of forecasting problem), where the neural network predicts a numerical outcome for some input data.

Now we have seen what a neural network's execution looks like; we've taken that plunge into the cold water. Next, we can ask the question, how can we use these tools to solve real-world problems? To answer it, we will step out of the pool and gradually ease back in with a couple of examples. First, we look at classification examples, and later we look at a regression example.

CLASSIFICATION USE CASES

Suppose we have the following task: a Hollywood production studio hires us to implement a tool that can scan movie reviews from users in some message forum and decide whether the reviews are positive or negative. This is a classification problem. We have some data—in this case, a few lines of text from a movie reviewer describing their thoughts on the movie—and we need to create a tool that can analyze that data and tell us whether it's a positive review or a negative review.

If you take a moment to think about this problem, you'll realize that it's not as simple as it might initially seem. Naively, we might suggest to simply look for some key words—*bad*, *good*, and *OK*—and decide that the review is either positive or negative based on the occurrence of those key words. If we take such a simplistic approach, we quickly realize that the problem is more nuanced than this. Consider the following sentences: "I was worried that this movie was going to be terrible. I was wrong." Clearly this is a positive review, but it has two negative words in it: *terrible* and *wrong*. Now consider these: "The greatest thing about this movie was when it ended. Even the best actors couldn't save it." This is an example of a negative review that has a few generally positive words in it: *greatest* and *best*. It's evident that we can't simply look at the words in isolation; we must consider the body of the text and the relationship each word has with every other word in the text.

These types of problems belong to the *natural language processing* (NLP) field of research, and neural networks are the best models we have for solving NLP problems today. When you ask Alexa to order that new pair of socks or tell Google Home to play a different song, an NLP neural network is processing your speech and making sense of your commands. So how do we solve the movie-review problem? How do we teach a neural network what makes a positive review?

The first thing we must do is get hold of a training data set. In this case, a training data set is simply a large collection of movie reviews that we can present to the neural network to train it on what positive and negative reviews look like. A training data set typically has thousands of samples; in our case, a sample is a single movie review. A training data set also has unique pieces of information called *labels*. Since we want to train our model to learn the meaning of positive and negative reviews by looking at a large number of training samples, the samples must contain classification labels that say "positive" or "negative" for each review. These labels are typically produced by humans—researchers and volunteers— who painstakingly label thousands of samples from training data sets.

We have been hired to create a tool that can classify movie reviews. We construct a neural network that can take bodies of text as inputs and then output a classification: positive or negative. We must find a training data set that we can use to train our newly created model. Thankfully, Stanford University publishes a data set of labeled movie reviews scraped from the IMDb (Internet Movie Database) website. The data set, which is freely available to everyone, contains 50,000 samples of movie reviews split into 25,000 training samples and 25,000 testing samples. Training data sets are typically split into training and testing samples. When we are training the neural network, we want it to learn from the training data, but we don't want it to "memorize" the training data. If the neural network learns the correct answer for each training sample by memorizing the training samples, that doesn't really tell us how well the network will perform in a real-world situation. The testing samples help us gauge whether the neural network has truly learned information that can be applied to unseen data or if it's simply memorizing our training data.

As we saw in our simple example above, neural networks are mathematical systems that work with numbers. They expect numbers as inputs, transform those numbers using mathematical operations, and output numbers. Part of our work in designing a neural network model to solve a problem is to figure out how we can represent, or *encode*, our data as numbers. Suppose we have the following review: "The movie was extremely good." How do we convert this sentence into a set of numbers? Computer scientists and statisticians have developed a few ways to achieve this.

One way is to first come up with a dictionary (or vocabulary) of words used by the reviews in the training data set. We can create a list of all words used by all the reviews in the data set. Then we identify the most frequently used words in the reviews. For example, suppose we produce a list of 80,000 words used by the movie reviews. We can select the 10,000 most frequently occurring words in the reviews and create a vocabulary of 10,000 words. Each word in this vocabulary is unique, meaning that it occurs only once in the list, and each word has an index associated with it. The first word has an index of 1, the second word has an index of 2, and so on. Let's recall the sentence we want to analyze: "The movie was extremely good." We can break the words into a vector: [The, movie, was, extremely, good]. Next, we check the index of each word in the list of 10,000 words and replace the words in the vector by their indices. But as we are doing this, we notice that the word *extremely* doesn't appear in the list of 10,000 words we have created. That means that the word *extremely* isn't used frequently enough in our training data set, and it is not part of our vocabulary, so we can discard it. The vector then becomes [The, movie, was, good]. When we use the indices of each word in the 10,000-word list, the vector becomes [10, 4, 22, 100]. We can interpret this vector as follows: the word *the* is the 10th word in the list, the word *movie* is 4th, the word *was* is 22nd, and the word *good* is 100th. Now we have managed to convert our review into a set of numbers, a necessary step because our neural network needs numbers to process information; we are well on our way to being able to ask our model to predict whether this review sample is positive or negative. But we can't do it just yet; we must make one more modification.

Since neural networks are dealing with numbers, we want the range of values for all inputs to be roughly the same; this is called *normalizing* the data. For example, suppose we have a set of reviews that happen to use words found toward the top of the vocabulary list. That is, all the words in these reviews have indices of less than 100. This would create a word vector where all values are less than 100. Then, suppose the next set of reviews comprises mostly words found in the bottom half of the 10,000-word list. Say the vectors consist of values that are greater than 8000 (e.g., [8001, 9245, 8444, 9001]). Recall the way that inputs are processed by the neurons in the network. We use the input values and perform a weighted sum operation over the neuron's connection weights, and the result of this operation gets modulated by the activation function of the neuron. High values and low values inputted into a neuron affect the output of that neuron differently. We don't want the output of the network to be skewed based on which portion of the list the review words came from, because that information may not have anything to do with the actual *sentiment* of the review. We want the network to learn what makes a review positive or negative. This requires the network to learn the semantic meaning of each word and the relationship between words in the vocabulary. The positions of the words within the vocabulary list, however, were arbitrarily chosen. When we created the vocabulary of 10,000 words, we did not organize the words in any meaningful way; therefore, we don't want the network to learn to weigh words differently based on the portion of the list they came from, which would certainly happen if some indices were much larger than others.

One way we can normalize the input vector so that the values for all words exist in the same range is known as *one-hot encoding*. That is, instead of the 4-dimension (four-valued) vector [10, 4, 22, 100] for "The movie was good," we create a 10,000-dimension vector, where we place a 1 at the index in the list for each word that appears in the review, and we place a 0 for words in the list that don't appear in the review. This would produce a vector of 10,000 values that looks like this: [0, 0, 0, 1, 0, 0, 0, 0, 0, 1, . . . (eleven 0s), 1, . . . (seventy-seven 0s), 1, 0, 0, 0, . . . (zeros all the way to the 10,000th place)]. It might seem complicated, but it really isn't. We have placed a 1 at the 4th place in the vector, a 1 at the 10th place in the vector, a 1 at the 22nd place, and a 1 at the 100th place. Everywhere else we have placed a 0. Now we have a method for encoding reviews as vectors of values that can be presented to a neural network for analysis. This vector encodes information about the words in our review but does not place more weight on a group of words versus another based on where in the list they appear. The vector is normalized, and all values are 0 or 1.

There is an advantage to using this one-hot encoding method, which is that all input vectors—that is, all reviews—will result in the same vector size. For example, we might have different reviews consisting of different lengths of text, but the process we are describing for generating the input vector will generate vectors that are all the same size: 10,000 values. As we will see, this is important because our neural network needs to know how many values are in the input, and the number of values cannot vary between samples.

The astute reader may be noticing a few potential problems with this method of encoding our input data. By choosing a vocabulary that is significantly smaller than the number of words in the body of the training data set (and in the English language in general), we have conceded that some words will be discarded from our reviews. These are the words that appear least frequently in the training data set. How do we know that we are not discarding valuable information? Furthermore, the one-hot encoding method we described has a flaw. Suppose that the word *good* appears three times in the review. With the proposed encoding method, we can only encode each appearing word once, as there is only one index for the word *good* in the list. Thankfully, the problem we are trying to solve is to classify the review as a negative or positive review. We are not trying to provide a degree of positivity or negativity (because it wasn't asked of us). For example, our model does not have to classify the review into "good," "great," "greatest," or "bad," "terrible," "worst." This means that losing such information as how many times the user said this movie was "good" may not be as important. But in general, these are all valid concerns. And there are other more complex methods of encoding information that avoid some of these issues and help encode word frequencies that would be necessary for multiclass classification, where we want to predict degrees of sentiment as opposed to a binary case of positive or negative. Even with the downsides we have discussed, however, this method still works pretty well and can produce really good results for binary classification tasks like the one we are discussing.

All right, so now we have our input vector. What do we do next? We present it to our neural network model. Let's define an architecture for a neural network model capable of solving this problem. We can design a neural network that has an input layer, two hidden layers, and an output layer. The input layer will accept 10,000 values. The first hidden layer will consist of 32 neurons with a ReLu activation function. The second hidden layer will consist of another 32 neurons also with a ReLu activation function. Can you guess how many neurons will be needed for the output layer? The output layer will consist of 1 neuron, and it will have a sigmoid activation function. It is important that we choose a sigmoid activation function for the output neuron because we want the output to be a probability. That is, we want the output to tell us the likelihood that the review is either a positive review or a negative review, and sigmoid functions output values in the range of 0 to 1, which can be interpreted as a probability. If this is not clear yet, don't worry; it should be made clear soon enough.

Let's discuss the architecture of the neural network. We are choosing the ReLu activation function for the neurons in the hidden layers. ReLu functions are simple: they take an input value; if the value is negative, they output a 0, and if the value is positive, they output the same value (i.e., the input value). The reason we are using the ReLu activation function is that, for most artificial neural network architectures and for most use cases today, researchers have empirically determined (by trial and error) that ReLu functions work best. For a long time, sigmoid functions were very popular, but ReLu functions have a property that makes the training process easier, so over time most researchers started adopting the ReLu function. The more interesting question is why ReLu functions work so well, and unfortunately the answer isn't exactly known. As the famous AI researcher Geoffrey Hinton once put it (and I am paraphrasing), "This is all made up."

Some of what we do in artificial neural networks is inspired by biological systems; as we know, biological neurons have internal states that cause them to output signals based on inputs but in a nonlinear way. That is, the strength of the output signal isn't always proportional to the strength of the input. ReLu, sigmoid, and other mathematical functions help add nonlinearity to artificial neurons, but that's where the similarities to biological systems stop. We do not yet have a deep enough understanding of biological systems to form a framework that allows us to find the appropriate activation function for an artificial neuron. Instead, researchers try different ideas empirically and keep the ones that work. Answering the question of how we choose the size of the hidden layers—how many neurons we want per layer—is also more of an art than a science. There isn't a set of rules that prescribe a specific number of neurons for a given problem. Over decades of research, some intuition has been built. There is a correlation between the size of the input, the size of the training data set, and the general range of neurons we should use, but no formula will tell us how to design the network. Again, much of this has been learned empirically. Researchers try different architectures and test their models, and then they adjust them when the systems don't work right away.

We have designed a neural network model and a process we can use to encode the input data into numbers that we can feed the neural network. We selected a sigmoid activation function for the output layer because we want to interpret the output as a probability. So how does that work exactly? How do we interpret the output of the neural network we have built? Recall that the goal of our neural network is to classify the input—a movie review—into one of two classes: positive or negative. As we know by now, neural networks only understand numbers. Therefore, we need to encode the classes "positive" and "negative" using numbers. We can do that by assigning to the class "positive" a label of 1 and assigning to the class "negative" a label of 0. We could certainly do the reverse: we could say the class "negative" is 1 and the class "positive" is 0; we just need to be consistent in our data set. When the neural network outputs a value that's between 0 and 1—because of the sigmoid function—we interpret that value as a confidence level that the input belongs to class 1 or 0. For example, suppose the network outputs 0.78, and we have established that a label of 1 means a positive review. This output tells us that the neural network is 78 percent confident that the review is positive. That is, based on the training data, there is a 78 percent probability that the review it's currently analyzing is a positive review.

Similarly, an output of 0.5 means there is a 50 percent chance of the review being positive.

We should take a moment to appreciate the importance of what we have accomplished. We have just taken an input sequence of text, converted it into a format that a mathematical model—the neural network—could analyze, determined how to design a neural network that can classify text into different categories by learning to pick up on emotional cues (this type of classification where we try to understand the mood of the author is also called *sentiment analysis*), and learned how to interpret the output of the neural network. We have seen that the process of solving this problem, while not intuitive, is surprisingly simple when we break down the set of operations. It's all just a set of multiplications and additions, with activation functions thrown in the mix.

The task of performing sentiment analysis was, for a long time, a very difficult problem to solve. As we saw, text isn't a simple sequence of words where every word has the same weight. Some words are more important for conveying the intent of a sentence than others, and words tend to have intricate relationships (e.g., a word close to the start of a sentence might emphasize the meaning of a word far away). Consider the sentence "It is interesting that in some countries, especially in tropical regions, rain falls for most of the year." The word "interesting" strongly relates to the word "rain" and the fact that it falls for "most of the year." Look at how many words we had to skip to get to the meaningful part of the sentence. Then consider the sentence "It is hard to explain what makes a movie interesting." In this case, the relationship is very close. The word "interesting" is directly related to its immediate neighbor, "movie." These nuances make it difficult to develop a set of hard rules in the form of "if this pattern, then that choice," which is how more classical intelligent systems were built. The reason neural networks are so powerful is that we don't need to explain to them what makes a review positive or negative—which is difficult because we have a hard time writing down general rules for those problems. (Give it a shot and see for yourself: try to write a set of rules for describing what makes a sentence positive, and try to apply those rules to a set of random posts.) Instead, neural networks learn to identify these rules for themselves. Take a minute and marvel at that!

As we were introducing the execution process of the neural network and learning to interpret the input and output, we skipped over a very important part: the training. In this chapter, we very briefly explain neural network training. Then, over the next two chapters, we gradually expand on our knowledge base and dive deeper into the training process. As we have seen, the execution process of a neural network is quite simple, and the mathematics used barely approach the high school level. The training process is a different story.

Why do we need training in the first place? In our examples thus far, we have used the weights of the connections between the network's neurons as if they had been magically preset to the appropriate values to provide the correct outputs. But unfortunately, this is not how neural networks begin their life. Initially, before the neural network is trained, the weights of the connections are randomly initialized, so the predicted output is largely inaccurate. It is only through the training process that the weights are adjusted until eventually the predicted output starts to make sense. The weights of the model represent the learned information. We start with a randomly initialized set of weights and a training data set. The data set is a collection of samples from the problem domain. For example, if we want to train neural networks to distinguish between apples and bananas, the data set must contain a diverse number of apple and banana images. If we want the neural network to distinguish between positive and negative reviews, we need the data set to contain samples of positive and negative reviews. During training, the neural network is presented with a sample from the training data set.

In an image classification example, a training sample is a single image from the data set; in a sentiment analysis example, such as our movie-review problem, a training sample is a single review from the data set. We present the training sample to the neural network, and it predicts an output. Because the neural network is not yet trained, the prediction is going to be inaccurate; for example, it might say that a highly positive review is negative. Since this is a training sample, the sample has a label that tells us its true class; the label will say "positive" for a positive review or "negative" for a negative review (remember: the labels are numbers, so 1 for "positive" and 0 for "negative"). We then use a mathematical formula to calculate how wrong the current predictions got it. That is, we calculate how far the predictions of the neural network are from the actual labels, and we use that calculation to adjust the weights of the model so that the next time the model sees this sample, it gets a little closer to the truth. During training, we perform this operation for each sample in the data set and for many cycles, or *epochs* (i.e., cycles where the neural network sees each sample once). The training process typically lasts many epochs, so the model sees each sample in the training data set multiple times, each time extracting a bit more information and transferring that knowledge to the network's weights.

Let's look at another classification example, this time from the vision domain. In the early 1990s, a significant amount of research went into creating algorithms for recognizing handwritten digits from U.S. mail envelopes. The training data set for this example consists of 50,000 handwritten digits. This classification problem has ten classes as we need the model to learn to distinguish between ten classes of digits: 0–9. The images of handwritten digits in this data set have a resolution of 28 × 28 pixels. In the sentiment analysis example, we converted the review samples into vectors where each word represented a different dimension of the vector. That is, for a review consisting of ten words, we produced a vector of 10 values, which we then converted into a one-hot encoding vector of 10,000 values. For image classification, each pixel is a different dimension of the input vector, so for images of 28 × 28 pixels, we produce a vector of 784 values (28 × 28 = 784), where each value represents the light intensity of the pixel: a value of 0 means a black pixel, a value of 1 means a white pixel, and any value in between represents a grayscale. Now that we have defined the input vector, we can proceed to design a neural network to classify handwritten digits.

We start with the input layer having 784 nodes, one for each pixel of the input vector. Next, we add a layer of 512 neurons with ReLu activation, followed by a layer of 64 neurons also with ReLu activation. Finally, we add an output layer of 10 neurons with a softmax regression function. The design choice of a 512-neuron layer and a 64-neuron layer is again driven by intuition and trial and error. Typically, we iterate over a few designs until we find one that's efficient (in terms of number of parameters) and performs well.

We have not discussed softmax regression functions yet. You can think of them as similar to a sigmoid function but for multiclass problems. Whereas a sigmoid activation function is used for binary classification problems with the intent of outputting a probability that the input belongs to a single class—cat vs. dog, apple vs. banana—softmax regression functions convert the output of the neural network into a probability where all outputs sum to 1. For example, our network has ten outputs, one for each possible class of the input digit: 0–9. For a given input image, the network might output the following ten values: 0.80, 0.1, 0.0125, 0.0125, 0.005, 0.0025, 0.0168, 0.0168, 0.017, 0.0169. This result tells us that the network is 80 percent confident that the input value represents a 0, 10 percent confident that it represents a 1, 1.25 percent confident that it represents a 2, and so on. If we sum all the output values together, we get 1. We can train this neural network on the 50,000 training images for five epochs. That is, the network will run through the training data set five times as it learns to predict the correct label for each image. After about five epochs of training, a neural network like this one can achieve around 98 percent accuracy in its predictions, meaning that the neural network correctly predicts the class of an input digit 98 percent of the time. This is quite remarkable considering that, before neural networks, no other image recognition algorithm came even close to that level of accuracy.

Why is an algorithm like this useful? In a practical sense, we can employ an algorithm like this at a sorting facility for the postal service. We might build an assembly line of sorting machines that receive a long stream of mail envelopes. The sorting machines can then look at the envelopes, recognize their zip codes, and sort the envelopes according to distribution regions. And there is another way that this algorithm is useful: it shows us that it's possible to build systems, with relative ease, that can interpret images. Artificial vision systems were considered among the most difficult areas of automation before neural networks, so developing an algorithm that could recognize an image and interpret it was a tremendous achievement. It gave researchers a sense that computer vision might yet be possible. We discuss more about computer vision and state-of-the-art vision algorithms in the next chapter, and in the process, we will discover a new neural network architecture that avoids a crucial problem with our humble MLP. For a 28 × 28 pixel image (a very small resolution, impractical for most real computer-vision use cases), we needed a neural network with 784 input neurons, which in turn set the stage for roughly the same number of neurons in the subsequent layers. Consider the size of a neural network required for resolutions of a few thousand pixels. It quickly becomes very difficult to process images using MLPs in this manner because we need larger MLPs with more and more neurons, which require dramatically more memory to process the input images. When we discuss computer vision, we will see a different type of network that manages to solve the vision problem with considerably fewer resources.

REGRESSION USE CASES

Classification problems, such as sentiment analysis and image recognition, are common use cases for neural networks. Another popular use case, regression problems, involves the prediction of a continuous value based on historical data. For example, we might have a data set of previous transactions and valuations for a particular stock trading in the stock market. We might want to create an algorithm that can analyze the market data and predict a future price for a given stock.

Or let's say that you are a project manager. You know your team's estimates for the number of hours required to complete a project are usually not very good: sometimes they overestimate and finish the project in half the time; sometimes they underestimate, and it takes much longer to complete than expected. Both underestimating and overestimating on projects contribute to the agelong struggle between management and developers. This is especially true in software development, where correctly estimating projects is notoriously difficult. It is obvious why underestimating the hours needed to complete a project is bad. And it might seem better to err on the side of overestimating, because if a project is finished in less time than quoted, then everyone should be happy, right? Unfortunately, it doesn't work that way. When a project is scheduled, a variety of resources are assigned to that project; if the project is then completed in significantly less time than was originally allocated for it, the effects ripple through other scheduled projects. Now management needs to figure out how to reassign the suddenly freed resources. It also probably means that the project was quoted to the customer at a much higher price point than it should have been: do this too many times in a row, and you start losing clients to competitors with lower price points.

It should be noted that the people doing the estimating are often experienced employees who are doing their best. But the reason estimating projects is so difficult is because of the number of variables that can affect the completion of the project. Sometimes requirements change halfway through the project. Sometimes, at the outset, the team does not have all the information needed to complete the project, and the information slowly trickles in as development progresses. In many cases, implementing a particular feature is contingent on a third-party company providing a key piece of equipment that inevitably gets delayed. These are all variables the person estimating a project must contend with and make assumptions about. Of course, you may say that all of this can be easily avoided by never starting any project until the team has absolutely everything they need. But this strategy would not be practical in dynamic environments like software development or any technology-based field. Innovation, almost by definition, requires uncertainty, and uncertainty means that engineers often can't predict all the problems they are going to encounter until they do encounter them, so estimating the time needed to fix the problems they haven't yet encountered is not an exact science.

One possible solution to this problem of project estimation is a regression algorithm. Provided that the management team has kept a database with descriptive features of past projects and the actual time it took to complete those projects, it might be possible to create a neural network that can analyze the data and learn to predict duration for similar future projects. If we recall from our previous classification examples, *features* are dimensions of the data. In the case of sentiment analysis, a sentence was converted into a vector of numbers, and each value in the vector was a feature of that vector. In the case of handwritten-digit recognition, the image was converted into a vector of pixel values, where each value was a feature of the image. In the project estimation example, a data sample would be a vector of features that describe a given project: number of engineers assigned, experience level of engineers, number of components this project depends on, time of year when the project is running, and so on. And the label to predict would be the estimated time—say, 160 hours. The difference between a neural network algorithm and a human estimator is that the neural network algorithm should be able to find patterns in the data and the relationships between the features better than a human could. A human might be inclined to put too much emphasis on a specific feature—for example, number of engineers assigned—and estimate the time to complete the work simply based on the number of engineers available. The neural network, however, through the training process should be able to pick out the predictive features for each individual project. In one case, the number of engineers may indeed be the driving force behind the work estimation, but in another, the relationship between other features might be more important. For example, the time of year when the project takes place and the client company may be correlated: at a particular company, the height of summer might be a time when most employees are on vacation, so getting vital feedback from them might be delayed. These nuances in the relationships among features that describe a project might be difficult for a human to pick up, especially when the project vectors contain many different features. But a neural network algorithm should pick up such details with relative ease.

Again, these types of forecasting algorithms are referred to as regression algorithms. The simplest form of regression algorithm is one that performs *linear regression*, which we discuss at great length in chapter 3. Linear regression is the process of trying to fit a line through a set of data points and using the line as an estimator for future points. What a neural network algorithm can do in a regression use case is automate the process of finding that *best-fit line* (fig. 1.22). We are now going to walk through a regression example and build a neural network estimator for a specific regression problem. Our Hollywood movie-review gig is done, and now we've been hired by a real-estate company. The company wants us to build a model to help valuate houses in a given city.

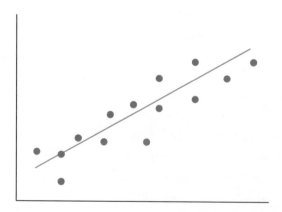

Figure 1.22 A best-fit line (*red line*) running through a set of data points (*blue dots*). We can use the best-fit line to predict the value for missing data samples.

By now you know the drill. The first thing we need is a data set of samples—vectors of data points that describe a house—with their correct valuations. There is a data set well known to machine-learning engineers: the Boston Housing Price data set, which describes a set of houses and their prices for different suburbs of Boston in the 1970s. Each of its 506 data samples consists of a vector of fourteen dimensions. That is, the data set comprises 506 houses, where each house is described by a list of fourteen features (see table 1.4).

Table 1.4 Feature Description of the Boston Housing Price Data Set

1	Per capita crime rate by town
2	Proportion of residential land zoned for lots over 25,000 sq. ft.
3	Proportion of nonretail business acres per town
4	Charles River dummy variable (1 if tract bounds river; 0 otherwise)
5	Nitric oxides concentration (parts per 10 million)
6	Average number of rooms per dwelling
7	Proportion of owner-occupied units built prior to 1940
8	Weighted distances to five Boston employment centers
9	Index of accessibility to radial highways
10	Full-value property-tax rate per $10,000
11	Pupil-teacher ratio by town
12	$B - 1000(Bk - 0.63)^2$ where Bk is the proportion of Black residents by town*
13	LSTAT: % lower status of the population
14	Median value of owner-occupied homes (in thousands of dollars)

*__Note:__ Row 12 bears addressing, but to not break the flow of the current explanation, it will be addressed immediately following the end of this chapter.

Each house in the data set is described by fourteen values referring to criteria that are suspected to correlate with the price of a house (e.g., per capita crime rate in the town, number of rooms in the house, etc.). The reason machine-learning algorithms such as neural networks are useful tools for analyzing this type of data is that it's very difficult for a human to look at each of those fourteen data points across hundreds of houses and pick out the features that are most predictive of the price of the house. Intuitively, we might guess that the number of rooms is an important one, but how important is it compared to the location of the house? And the pupil-teacher ratio in the area? The purpose of the neural network is to learn these relationships from the data.

We can create a neural network for the present problem as follows: The first layer contains 14 input nodes, one for each dimension of the data. The next layer contains 64 neurons with ReLu activation, followed by another 64-neuron layer. The next and last layer consists of a single linear neuron. A linear neuron is simply a neuron that calculates its output based on the weighted sum of the input and its connection weights without any further processing by an activation function (i.e., no threshold- or activation-related modulating). Recall that each neuron in each layer is connected to every other neuron in the following layer. The layout of the network is, as previously discussed, fine-tuned through trial and error. All we know for certain is that the first layer should contain 14 input nodes and the last layer should contain 1 output neuron. We need 14 input nodes because our input sample consists of a 14-dimension array. Similarly, we need 1 output neuron because all we want the neural network to do is output the predicted price of the house; that is, we just need a single number to be output. Instead of two hidden layers with 64 neurons each, we could have chosen two hidden layers with 128 neurons each or one layer with 128 neurons and the next one with 64 neurons. We could have also chosen three hidden layers with 32 neurons each, and in all cases the neural network would have learned to predict a price for the house that represents, with varying degrees of accuracy, the training data. Typically, we test different arrangements and choose the one that performs best.

Once we create a neural network, the process for training it on a regression problem is similar to the process for training it on a classification problem. The training data set, which in our case has 506 houses, also contains the target prediction for each sample in the data set; in other words, for each vector of fourteen values describing each house in the data set, we also have the true price for that house. The training is split into many epochs (i.e., cycles where the neural network sees all the samples in the data set at least once). At the start of training, during the first epoch, the neural network's parameters are randomly initialized, so the price prediction for each house is not very accurate. Thankfully, we have the true price of the houses, so we can measure how far off the network's predictions are from the true value of the house. Using this information (and a lot of calculus), we update the neural network's parameters to minimize the difference between the expected value and the predicted value next time through. Just as we saw in the classification use case, over many epochs of training, the neural network parameters (the connection weights) are updated and tuned to predict the value of a house given a set of descriptors (the features) for the house. The hope is that once the neural network is trained, it has learned the relationship between the features describing the house and the value of the house. Then we can show it a new house that is not part of the training data set, and the neural network should be able to predict the value of that house.

VECTORS AND VECTOR SPACES

We have now spent a bit of time discussing fully connected neural networks and how we can build them to solve classification and regression problems, the two most common types of problems in artificial intelligence. We know how to build a network of neurons, a layer at time, to accept an input vector and output either a single value or some prediction vector. We have also said in passing that a classification algorithm (which includes neural networks) can be described as a line that separates graphed points in N-dimensional space. The points on one side of the line belong to one class, while the points on the other side belong to the other class. Similarly, we have said that regression neural networks can be thought of as a best-fit line through a set of data points, where the line itself is a continuous predictor for all data points in that N-dimensional space. The problem with this explanation is that it's not easy to visualize.

Suppose you want to differentiate between images of a handwritten digit 1 and a handwritten digit 2. This "line separating points" analogy surely can't apply here, right? How can you think of images as "points," and then how can you use a line to separate images? It turns out that you can indeed treat images—or any type of data samples, including our houses and movie reviews—as points in some space. Typically, this space is multidimensional, so we call it a *hyperspace*. And the neural network is truly trying to find a line that separates these data points; however, because we are dealing with many dimensions, instead of a line (which separates points in two dimensions), we are trying to find a *hyperplane*. To understand this process, let's think back to our high school days and recall what a vector is.

In 2D space, a vector can be described as follows: $V = (X_1, X_2)$, where X_1 denotes the component of the vector in one dimension and X_2 denotes the component of the vector in the other dimension. A vector in 3D space can be described as follows: $V = (X_1, X_2, X_3)$, where again X_1, X_2, and X_3 describe the components of the vector in each of the three dimensions. We can close our eyes and visualize a point in 2D space with an arrow starting at the origin and ending at the point. This is a 2D vector. We can also visualize a vector in 3D space as a point floating somewhere in the 3D world with an arrow starting at the origin and ending at the point (fig. 1.23). But can you visualize a vector in 4D space? Unfortunately, we can't visualize spatial relationships beyond three dimensions. But it turns out that the mathematics of vectors in four dimensions are the same as the mathematics of vectors in two and three dimensions. Although we can no longer visualize it, a 4D vector mathematically looks like this: $V = (X_1, X_2, X_3, X_4)$, just like a 3D vector but with an extra dimension. Similarly, a 5D vector looks like this: V = $(X_1, X_2, X_3, X_4, X_5)$. And an ND vector looks like this: V = $(X_1, X_2, X_3, \ldots, X_N)$. So manipulating vectors in a multidimensional space is similar to the process of manipulating vectors in 2D and 3D spaces, which are much more intuitive to us. We simply have to account for the extra dimensions.

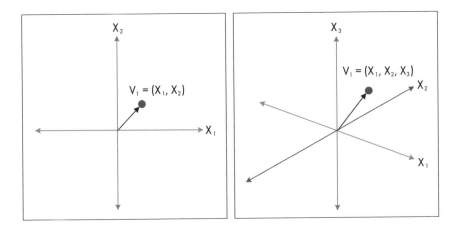

Figure 1.23 A vector in 2D space (*left*) and a vector in 3D space (*right*). A vector is just a point in some space with an arrow running from the origin to the point.

How is this useful? Once again, let us consider a 28 × 28 pixel image of a handwritten digit 1. To process this image through our neural network, the first thing we did was turn the image into a vector. We interpreted the color of each pixel as a number and then created a vector of 784 (28 × 28) numbers. That is, we created a vector V = $(X_1, X_2, X_3, \ldots, X_{784})$ of 784 dimensions. Something very subtle and extremely important is happening here. Conceptually, we went from interpreting this image as a collection of 784 pixels of independent colors to interpreting this image as a single point in a 784-dimensional space. This constitutes a very important assumption. We are saying that the 784 pixels are not just randomly independent pixel values that happened to resemble an image of a 1. Instead, we are saying that this image exists as a single point in a universe with 784 dimensions where each pixel value constitutes a component of the vector in each of the 784 dimensions. If we consider other samples of handwritten 1s, it turns out that these are all single points in a 784-dimension virtual universe as well.

And now for the really interesting bit. Vectors prove to have a special property. We know that similar vectors point in roughly the same direction, whereas vectors that are different point in different directions. Using our intuition from 2D and 3D experiences, we could imagine that in a 784-dimension world there is also an origin point, and from this origin point, we could imagine that we have arrows going all the way out to each of the points that constitute our handwritten 1s. Now, for fun, let's throw in images of the digit 5. These are also 28 × 28 pixel images, so they exist in the same 784-dimensional universe as our digit 1 images, but since they are a different digit, what would you expect this to mean in terms of vector space? If you answered that the digit 5 vectors would be pointing in different directions compared to the digit 1 images, you are correct. You deserve a break: go get a cold one! If you didn't, then perhaps go back to the start of this section and give it another try. This is an important concept that is central to machine learning and artificial intelligence.

Interpreting data samples as single vectors in some *N*-dimensional space means that you can imagine samples of the same class to be grouped somewhat together in that space and samples of a different class to be grouped in a different direction. Now that we have a way of interpreting data as groups of points, a classification algorithm simply needs to find a hyperplane (remember: a line in 2D, a plane in 3D, and a hyperplane in multidimensions) that bisects the vector space so that points on one side of the plane belong to one class and points on the other side belong to a different class (fig. 1.24). This is why vectors are so useful. They help us interpret our data in a way that lets us apply concepts from algebra and geometry to discover vital relationships among our data samples.

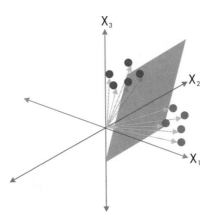

Figure 1.24 Two classes of vectors in a 3D space separated by a hyperplane. The vectors to the left of the plane (pointing to the *purple points*) belong to one class, and the vectors to the right of the plane (pointing to the *green points*) belong to a different class. Why is this useful? If we get a new sample and we do not know what class it belongs to, all we need to do is interpret it as a vector and see on which side of the plane it falls; then we can predict its class.

○ ● ○

In this chapter, our goal was to gain some understanding of how artificial neural networks work. It's hard not to attribute mystical qualities to these analytic tools, given their name. But as we are beginning to see, they are simply mathematical constructs consisting of simple operations, mostly multiplications and additions that are organized in sequential steps. (The sequential steps, of course, are the layers of neurons stacked one after the other.) We started the chapter learning about the history of artificial neural networks and the first artificial neuron, the McCulloch-Pitts neuron. We saw how the artificial neuron was inspired by our understanding of the brain and how biological neurons process input signals. The McCulloch-Pitts neuron, as the first incarnation of an artificial neuron, was a simple construct. It performed the minimum possible duties to be considered a neuron at all. As our understanding of biology progressed, helped by research from scientists like Donald Hebb, we learned that all inputs are not treated equally. There is a system of weights that can be adjusted to emphasize the signals from some inputs over others. This research from the field of biology made its way into the computer science field (although, at the time, it wasn't yet called computer science), and the artificial neuron was updated to include weights. This development led us to Rosenblatt's perceptron and the first neural networks that could do something useful, ADALINE and MADALINE.

We also saw that the first architectures of neural networks—namely, single-layer perceptrons—had a significant drawback. They could only be used to classify data points that are linearly separable. Now we know what that means because we discussed analyzing our data samples (images, text) in terms of points (vectors) in a multidimensional space. A data set that is linearly separable simply means that the data set's classes can be separated by bisecting the space in which they live using a line or some hyperplane. Unfortunately, in practice most classification problems are not linearly separable in their natural space. For example, if we wanted to classify images of cats and dogs and we used a linear classifier directly on the images of cats and dogs, we would find that our classifier would perform poorly. The raw vectors—in other words, the vectors produced by converting the images into arrays of pixel values—are vectors that usually contain a lot of data that's irrelevant to the problem (e.g., pixels of background information). Given these extra pixels and noise in the natural images, it's difficult to expect that as vectors they would live in a space that's neatly organized with cats on one side, dogs on the other side, and our separating hyperplane in the middle. Neural networks and the addition of extra layers (hence *multilayer* perceptrons) help us by transforming input vectors into a different hyperspace, where the resulting vectors *are* linearly separable. This is the secret of neural networks! They learn to transform input vectors into a hyperspace—in other words, into a different dimensional space—where the vectors can be linearly separated.

In the next chapter, we look at computer vision, a bit of its history, and its similarities to biological vision. We also introduce a type of neural network architecture, the *convolutional neural network* (CNN), that is the bona fide neural network architecture for computer vision.

But before we move on to the next chapter, we want to take a moment to address an important issue.

BREAKPOINT: CONFRONTING A TROUBLING LEGACY

Row 12 of the Boston Housing Price data set describes the "proportion of Black residents by town." This data set was compiled in the 1970s, and this feature (dimension of the data sample) overtly illustrates the prejudices of the times. Why are we including such a data set in an example in this book? This data set is one of the many "toy" data sets available online for machine-learning and artificial intelligence research. From tutorials and deep-learning books to online competitions, the Boston Housing Price data set is often used to benchmark AI models and compare the performance of new models to state-of-the-art ones. Yet, as ubiquitous as this data set is in the AI community, row 12 is almost never discussed.

We could postulate many reasons for why it is never discussed, but I suspect the answer is simple and should be of great concern. Mostly, people don't spend enough time looking at the data to even notice. The problem with row 12 should be obvious. It implies that there is a relationship between the number of Black people in a town and the value of the houses. The apologist's response is equally obvious: What's wrong with putting the information in the system? If it turns out that the model finds a relationship between Black people and house prices, how is that my fault? There are many problems with this interpretation.

Take a look at table 1.4 again, and review all fourteen features for the housing price data set. Whenever we create a training data set, there is a subtle danger of creating a *biased* data set. I would say that a training data set is always biased in some way, a shortcoming that is almost impossible to avoid. It's biased because someone constructed the data set by selecting a set of samples to be used for training the model. The fact that some samples are selected and some samples are not means that we are introducing bias into the system. Think of the handwritten-digits data set. It contained 50,000 training samples. We don't know how many individuals contributed handwritten digits to the data set, but suppose 5,000 individuals contributed ten digits each. In a population of 300 million, it is not clear that the 5,000 selected individuals' writing represents the way most people write numbers. So, by creating a data set of samples using digits from only 5,000 individuals, it's possible that we are training a model with some bias toward how those 5,000 individuals write, which does not translate well to the general population. Clearly, there are different levels of danger in bias. If the handwritten-digits data set is indeed biased toward how a few people write digits, the worst that could happen is the model won't generalize well in production, and it will make mistakes in interpreting the handwriting of the general population.

Now let's go back to table 1.4. Right away we can see that there is stronger bias in this data set, even if we disregard row 12. The data set consists of fourteen features that are used to predict the price of a house. By the mere act of selection, we have already placed more emphasis on those fourteen features than on any other possible predictors for house prices. Now consider row 12. When we include the number of Black people as a possible predictor for house prices, we are making the data set biased toward Black people—regardless of whether the data shows a positive or negative correlation between the feature and the house price. But as opposed to the handwritten-digits example, this bias has terrible consequences. There are three different possibilities for how row 12 can affect the data: Black people contribute to an increase in house prices; Black people contribute to a decline in house prices; Black people are not a predictor for house prices. Only one out of the three possible outcomes is positive toward Black people. This means that right away there is less of a chance for this data set to benefit Black people, and the fact that they were selected as a feature in the data set puts them, unfairly, in a defensive position at a simulated trial.

Suppose that, indeed, our model finds a relationship between the number of Black people and house prices, and suppose that it's a negative relationship. What the model cannot do is explain the reason for this relationship. Does the result mean that if Black people move into an up-and-coming neighborhood, the real-estate value will drop? Or does it mean that, because of the well-documented "racial wealth gap,"[8] Black households have lower median net worth and can often only afford to live in areas where real-estate value is already low? It is all too easy for social biases and prejudices to skew an analyst's interpretation of the results and lead them to conclude, "Black people are bad for the real-estate value in a region." Such an interpretation can then give rise to policies of segregation, where townships might decide to disallow Black people to move in, lest their real estate depreciate. This further alienates and disenfranchises people.

The second problem with this example, if the first one wasn't troubling enough, is the statistical significance of the data. The data set has 506 sample houses. Do we know how those samples are distributed around Boston? Can we draw conclusions about the relationship between the input into our model and the results without knowing this information? Yet it would be all too easy to use this result as a blanket reason to support prejudicial policies. Suppose it turns out that there is a negative relationship between Black people and house pricing in this data set. What about a different data set? What if the data set was larger and included other regions where Black households' median incomes are higher? Or what if the data set included other ethnicities: Do we know how good the Irish or Italians are for real-estate value? How about Jews? We all know how this ends.

This data set was generated in the 1970s, and it is, by definition, racist. Yet many people use it today without realizing its racist component, and that's an important danger we want to address. Today, we have more access to data than humans have had in the history of our civilization. And now, for the first time, it appears that we also have the

---

8. Vanessa Williamson, "Closing the Racial Wealth Gap Requires Heavy, Progressive Taxation of Wealth," Brookings Institution, Dec. 9, 2020, https://www.brookings.edu/research/closing-the-racial-wealth-gap-requires-heavy-progressive-taxation-of-wealth/.

power to modify it, transform it, and extract patterns and information from it. But it is not clear that we have the ability to understand our results and make sense of the data. How do I know that people use the data without paying enough attention to it? In several online tutorials and at least one book on deep learning, the Boston Housing Price data set is used but not really explained. In most cases, only the first few rows (rows 1–3) are listed with explanations of what the features mean; the rest of the rows aren't explained. Instead, they just show examples of the vectors of fourteen values. And very little emphasis is placed on understanding what those values mean.

Was this done on purpose, to avoid an embarrassing part of our history? A yes answer to this question would be bad, but my guess is the answer is no, and unfortunately this answer isn't much better. It shows our willingness to use data without really understanding it and that we expect our AI models to be "intelligent" enough to avoid biases in the data. But as we have seen in this chapter, the "intelligence" in AI is granted too quickly. Our models simply learn patterns from the data. If we produce and input data that is biased, our results will be biased—and we always produce biased data!

There is a real danger in deciding that because we can't understand our data, we need to use an AI model to make sense of it. If we blindly shovel information into a model and make policies based on the results, we are fully answerable for the biases that are perpetuated. The responsible approach is to first understand the implications of our data, especially the biases buried in it, and then use algorithms to find patterns that we humans have trouble discerning.

2

HELLO, PANDA!

We are well on our way to having a basic understanding of neural networks and how they function. In the previous chapter, we were introduced to the MLP (multilayer perceptron). We learned a bit about the history of the artificial neural network, which helped us grasp why neural networks work and what inspired their creation and evolution. The MLP is a powerful general approximation function, and it can be used to analyze data for a wide array of use cases, from classification problems to different forms of regression. In this chapter, we discuss a different architecture of neural networks called the convolutional neural network (CNN)—the star of image-processing algorithms and the backbone of most computer-vision systems today. More importantly, MLPs and CNNs are the building blocks for the neural network architectures predominately in current use. If we can understand MLPs and CNNs, we will have a solid foundation for understanding what is happening inside most neural network implementations for the foreseeable future. We have already seen MLPs; let's get started on CNNs.

One of the classical use cases for artificial intelligence is computer vision—creating an algorithm that allows a computer to perceive the world around it through a camera lens. The development of computer vision brings with it varied benefits and applications: security systems that monitor live video footage for suspicious activity, manufacturing assembly lines that ensure product quality, and so on. It allows us to build autonomous systems that can navigate complex environments without bumping into obstacles; think of self-driving cars, pilotless airplanes, or rescue robots that can risk harsh conditions and move more quickly than humans to reach disaster victims. Computer vision may allow us to build medical infrastructure for automating patient diagnosis by having a computer analyze large volumes of pathology images and thus accelerate the diagnosis process for the long line of anxiously waiting patients in overwhelmed health care systems. Some researchers believe that in the future we will have companion robots that help take care of individuals who live alone. Computer vision can help these robots understand their human companions by following visual cues and offering the necessary support.

There are many practical benefits to computer vision, and we'll continue to examine some of those. But besides that, we spend this chapter discussing vision systems because they are so incredibly interesting! There is something about vision that fires up our imagination. Although vision is quite common in many biological systems, there is still a lot that we don't know about it. The human vision system is probably the most complicated vision system ever produced, and although researchers have been successful in unveiling some of its mysteries over the past few decades, much of it remains hidden in the brain's visual cortex like a wonderfully kept secret. This combination of complexity and the unknown makes it a fascinating subject, whether we're studying biological or artificial systems, and as we will see, as we slowly chip away at the unknown and unveil the hidden machinery, we often find beauty and elegance (with a little bit of chaos sprinkled in for good measure).

We begin our discussion of computer vision by breaking down one of the most successful artificial vision algorithms ever created by humans: the convolutional neural network. We will peer into each layer of the CNN and inspect what operations are taking place inside. This process will help us demystify further the algorithmic aspect of the neural network. We will see that although it bears a complicated name, it consists of simple and systematic operations. By the time we are done discussing vision, we will have laid the foundation for understanding, at least at a surface level, what makes artificial neural networks powerful and will start getting an inkling for their limitations, as well. We will compare computer vision and biological vision and explore a fascinating phenomenon that happens with artificial neural networks. It turns out that our artificial neural networks exhibit many properties that are also found in some biological visual systems, but incredibly, these properties were not purposefully built into the artificial neural network by the designer. These are what we call *emergent properties*, which arise naturally in complex systems and might be a side effect of stumbling onto fundamental truths (more on this later). By following the thread of these emergent properties, we are going to discuss the similarities and differences between biological and computer vision as it exists today.

Although the need for computer vision is well established, and its use has become more ubiquitous in recent years, it wasn't always obvious that we could create a set of discrete steps that would let a computer detect objects in an image (those steps are essentially what a CNN is). In fact, for a long time, this seemed like an intractable problem. It may be difficult for us to realize how truly difficult a problem computer vision is—because we are used to seeing and recognizing objects! Distinguishing between different objects is something we can do effortlessly. But let's step back briefly and think about how a computer might see the world. In our human eyes, vision begins with photons of light entering the cornea. In computer systems, vision begins with pixels.

A camera snaps a picture of the world, and the picture is a 2D image. The computer sees a 2D image as a group of pixels. Let's consider an image of a cat (the "cat" has become the "Hello, world!" of computer-vision examples). Setting aside computers for a minute, suppose we meet someone from a different planet, and we want to explain to them what a cat looks like. First, we show them a picture of a cat lying on a bed. In this picture, the cat is prominent; it takes up a large portion of the image. It is facing the camera with two perfectly triangular ears pointing up, symmetrically placed on either side of the head. The cat has golden fur with dark stripes running the length of its back. Let's further assume that this extraterrestrial (ET) creature understands things in a quite literal way. To her, "cat" now means an object that has ears that are triangular, perfectly symmetric, and pointing straight up. She is building a set of simple rules that define "cat" for her. The fur of a cat must be golden and striped black. Now we show a second image of the same cat to our ET friend. In this picture, the cat is leaping in the air with its tail pointing up. The head is slightly turned so that only one ear is partially visible. With this new image, our ET friend is lost! She does not know what she is looking at. First of all, there is a tail in this picture! The picture that we used to train her on what a cat looks like had no tail because in that image the tail was hidden. Also, in this image the cat only appears to have one ear, and it's not perfectly triangular and facing the camera.

If this sounds like a ridiculous example to you, it is because, again, you are used to interpreting objects visually. But the ET in our example does not have the same well-developed visual capabilities that humans do. Instead, she is building simple rules based on the one example we showed her and the fact that we told her that that's what a cat looks like. She does not have the ability to extrapolate from a single cat in one position to what that cat might look like in different positions and know that it is still the same cat. This is exactly the problem that computer vision presents. The computer is an infuriatingly literal friend. We start with an image and must build a set of rules, or check marks, that a computer can use to determine whether the object in the image is a cat. In fact, the computer-vision problem is even larger. You see, I did a quick sleight of hand with my ET example. To simplify my example, I gave the ET powers to differentiate between the object we wanted to identify (the cat) and the background (the bed). But with computer vision, the computer doesn't even know where the foreground ends and the background begins!

We can generalize on the previous cat example by referring to the problem of identifying the cat in different positions as the *class invariance problem*. A computer-vision algorithm, if it is to be useful, must be able to recognize different classes of objects while allowing for each class sample to look different, to be of different sizes, and to be seen from different viewpoints and at different depths in the camera's field of view. Consider different breeds of cats: Persian, Maine coon, British shorthair, American shorthair, ragdoll, sphynx. An algorithm trained to recognize "cats" must be able to recognize a picture of a cat even when there is wide variability in how cats of different breeds look and how the same cat might look from different viewing angles.

TACKLING THE COMPUTER-VISION PROBLEM: TOP-DOWN APPROACH

When considering the problem of building a vision system, there are two different approaches we could follow. In the top-down approach, we start by asking ourselves what the elemental qualities common to all cats are (cats or elephants or camels—whatever it is that we are trying to identify). Once we identify the visual cues that signal "cat" to us, we can then proceed to construct a set of rules we can use to teach an artificial system to identify cats, so long as the images it is inspecting exhibit the qualities that we have associated with cats. In artificial intelligence, we call these qualities features (sound familiar?).

We can think of classical artificial intelligence algorithms as decision trees following a hierarchy of rules: if this happens, then that happens, otherwise something else happens; and from these two branches, subsequent rules and decisions can follow. Now consider building a computer-vision algorithm to identify cats in an image using our top-down approach. We start by writing down a set of simple rules that describe a cat. First, we will probably describe the edges of the cat—in other words, the boundary between the cat and the background. We will definitely note the triangular shape of the ears, which are a very prominent cat feature. Our rules will describe a set of edges that form triangles without a base, like this ^ ^. We might create different versions of these edges to cover multiple viewing angles. Most cats also have tails, so our rules must include edges that follow the shape of tails. We will continue this process and describe the torso, the legs and feet, and the iconic sharp ellipses of the eyes. Once we have constructed enough of these rules, we can build an algorithm that analyzes images and tries to find as many of these edges as can be found in the image. If enough edges are found, we might conclude that we have a cat in the image.

I have just described a very simple algorithm using a concept known as *feature engineering*. Feature engineering is the process of manually either selecting or constructing a set of features that describes the class of objects we want to identify. We construct a list of features and proceed on a feature-matching expedition through our data set. In each data sample (each image, in this case) we try to match as many of our engineered features as possible, and if enough features are matched, we assume we have found our object. I say "if enough features are matched" because we need to use a threshold of features that we want to match before declaring an object found. If the threshold is too low and we return a match after a single feature has been found, we might realize that our algorithm is very noisy and would return many erroneous matches.

For example, suppose that in our list of relevant cat features, we have the triangular ears ^ ^. If our algorithm predicts "cat" for every image that contains that feature without considering other features, we may be in for a few surprises as that shape is not found exclusively in cats. For example, the serrated edge of a handsaw has similar shapes; so do the shingles of some rooftops. To determine that a specific object has been found in an image, our algorithm has to find a collection of features belonging in our object's class. The number of features that must comfortably be matched to achieve an acceptable success rate is the threshold that we described. A low threshold results in many false positives—incorrect predictions for cats—while a high threshold might result in an algorithm that is biased toward the features we engineered and may miss many examples of cats where all their features don't exactly match our definitions (false negatives).

Feature engineering was one of the earliest techniques used in artificial intelligence algorithms, and it is still used in some cases today as it is much simpler and computationally less intensive than our current state-of-the-art approaches. Face-tracking algorithms that put a green bounding box around faces in the viewing screen of digital cameras still use feature engineering. They use simple features describing the shapes of eyes, noses, and the average distance between the eyes and the nose of most individuals. Wherever those features are matched in an image, they calculate a bounding box around the face. The downside of feature engineering is that it is very time-consuming and requires very experienced and well-trained engineers to design features that are specific enough to properly describe the subject but generic enough to allow for class invariance (the features must work for different shapes of faces and at a range of depths from the lens). This means that only a small list of the most prominent features can be created, which leads to algorithms that can work quite well in some controlled scenarios: for example, face detection on a camera where you know the subjects will typically be facing the lens head-on and at a reasonably predictable depth. But in general settings where the conditions aren't controlled, these algorithms prove too rigid for their predictions to be trusted. Lastly, the biggest limitation to pure feature engineering, and which binds it forever to specific and controlled environments, is that for every new class of items we want the algorithm to identify, we must start from scratch by designing new features for the new class of objects. This makes this class of algorithms very hard to scale to complex environments. Consider the difficulty in constructing a vision system for a self-driving car if we must handcraft the features of each obstacle a car might encounter in its lifetime (the roads, every type of vehicle, pedestrians, signs, trees, animals, random objects on the road, etc.). It would be impossible.

TACKLING THE COMPUTER-VISION PROBLEM: BOTTOM-UP APPROACH

The second approach to artificial vision systems is more closely related to biological vision systems. This is what we call the bottom-up approach. A biological vision system—specifically, the mammalian vision system—starts with light entering the eye. As different photons enter the eye and hit specific parts of the retina, the eye and the brain begin to construct a hierarchy of features that eventually leads to our understanding of the world we see. In computer vision, these points of light are the pixels that make up an image. Our goal is to devise a method by which our algorithm can construct a set of features by starting from the pixels. Ideally, the learning process of our algorithm will be generic so that for any class of objects, the learning process will begin with the pixel and proceed from there. Intuitively this seems like a less restrictive approach compared to feature engineering. With the bottom-up approach, we do not decide which features are important for any class of objects. We let the algorithm figure that out on its own. This means that the algorithm is free to see beyond our human box and discover its own way of seeing the world. As we let the algorithm itself learn to see, we might, paradoxically, learn more about our own vision than if we injected the algorithm full of our biases. An important aspect to note of feature engineering is that every engineer will design different features for describing the same thing. Shouldn't features that describe the world be determined by the data itself and not by the subjective judgment of an individual? Shouldn't vision rely on a set of fundamental truths about the world? The bottom-up approach attempts to solve this problem by removing the human from the feature-selection process.

Let's recall our cat-recognition experiment. If we wanted to implement a systematic way of extracting the outline information of a cat, how would we go about doing it? It turns out there are mathematical operations that can do exactly this. These fall under the *image processing* area of computer science. We can take an image of a cat and generate a new image containing only the outline of the cat. That is, we can remove all the extra information depicting the cat (color, fur texture, etc.) and leave just the edge information, simply by using math. For each pixel in the cat image, we update the pixel value based on a special 3 × 3 filter matrix. The process of updating the pixel values in this manner can result in a new image containing just edge information. If this is not clear, don't panic; we will explain it with a detailed example.

These image-processing operations are called *convolutions*,[9] or *filtering operations*. To carry out convolutions, we start with an image. In image processing (and computer graphics in general), we need to interpret the image as a mathematical construct. In other words, we need to figure out a way to interpret the image as a collection of numbers on which we can operate. Images are organized as arrays of pixels. That is, an image consists of a two-dimensional surface with a resolution of N-width × M-height pixels. An image that is 100 × 100 pixels has 100 pixels per row and 100 rows of pixels. Each pixel in the image has a value associated with it, and this value ranges between 0 and 255. If the image is black-and-white, we say that the image only has one color channel. For single-channel images, each pixel in the image has a single 0–255 value associated with it. This value describes the light intensity of the pixel. If the value is 0, the pixel is off and appears as black. If the value is 255, the pixel is fully lit and appears as white. For any value between 0 and 255, the pixel appears as a shade of gray.

If, instead of a black-and-white image, we are dealing with a color image, we say that the image has three color channels: red, green, and blue. In this case, each pixel has a 0–255 value that defines the color intensity for each channel. For our purposes, we will assume that we are working with black-and-white images because it makes the examples easier. But note that for color images, the process is the

---

9. While often referred to as "convolutions," these operations technically are correlation operations. A true convolution operation requires flipping the filter kernel first.

same. The only difference is that the operations are performed for each color channel.

Now that we understand how images can be interpreted mathematically, let's consider a black-and-white image of a cat, and let's assume the image is 100 × 100 pixels. We will interpret the image as an array of 100 × 100 values. Next, we perform an operation that's known as *edge detection*. This is a well-known image-processing operation that takes an input image and transforms it into an image of the same dimension where all pixels except for the edges of objects have been turned off—that is, they appear black (fig. 2.1, center panel). To extract edge information from an image, we take a specially constructed 3 × 3 matrix called a *filter*, or *convolution kernel* (3 × 3 is a common size, but larger sizes, such as 5 × 5, 7 × 7, and so on, may also be used), and use the edge detection filter matrix described below. We overlay the center cell of the matrix over a pixel in the input image and calculate a weighted sum across the pixels covered by the filter matrix. This becomes the new value for the pixel in the output image. If we perform this operation over every pixel in the input image, the result is a new image showing only the edges of the figures in the original image (fig. 2.1, center panel).

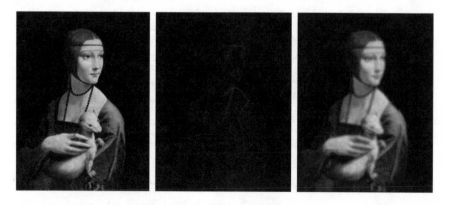

Figure 2.1 *Left*, an input image, Leonardo da Vinci's *Lady with an Ermine; center*, the result of an edge detection operation; *right*, the result of a Gaussian filter operation. Both operations were performed following the process described in figure 2.2.

Now let's look at figure 2.2 and see in more detail what we mean by a weighted sum using the 3 × 3 filter matrix. To perform edge detection on this 12 × 12 image, we use the edge detection filter matrix. We overlay the filter matrix over the top left 3 × 3 corner of the image and perform the following operation: $1*(-1) + 3*(-1) + 10*(-1) + 18*(-1) + 88*(8) + 43*(-1) + 6*(-1) + 2*(-1) + 55*(-1) = 566$. This gives us the value of the first pixel of the new image. Since this value is 566 and our pixel-value range is 0–255, we clamp the value to 255.[10] Next, we slide the 3 × 3 filter matrix to the right by one pixel and perform the same operation again: $3*(-1) + 10*(-1) + 0*(-1) + 88*(-1) + 43*(8) + 23*(-1) + 2*(-1) +55*(-1) + 65*(-1) = 98$.

This gives us the pixel value of the second pixel in the new image. We can continue to slide the filter matrix this way until we reach the right edge of the image. We then move the filter matrix one row down and start from the leftmost pixel again. By the time we reach the bottom right corner of the input image, we will have produced an output image showing only the edge information of the original image. At this point, our resulting image will contain only the edge information of the input image. If instead of edge detection, we wish to perform a blur operation, we can use the Gaussian blur filter in the same manner and then divide each resulting value by 16:

$$1*1 + 3*2 + 10*1 + 18*2 + 88*4 + 43*2 + 6*1 + 2*2 + 55*1 = 556$$

$$556 / 16 = 34.75$$

In this case, we round up, and the first pixel in the new image carries a value of 35. (In the next section, we discuss the importance of the values in the filter matrices.)

---

10. Clamping in this context means maintaining the value within a specific min-max range. In this case, our range is 0–255.

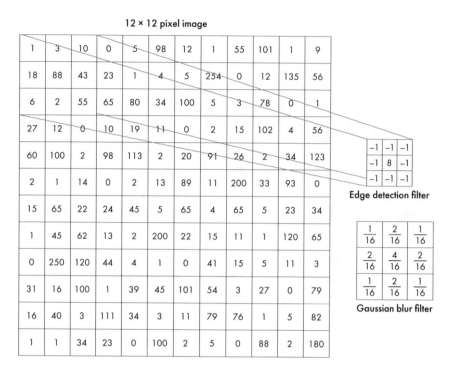

Figure 2.2 *Left,* a 12 × 12 pixel image with the associated value for each pixel; *above right,* an edge detection filter matrix; *below right,* a Gaussian blur filter matrix.

Isn't this amazing? By simply performing a set of simple multiplications and additions between a small matrix and the pixel values in an image, we can remove most of the information from the image and leave just the edge information! Furthermore, if we plug another set of specially prepared values into the filter matrix, we can produce a blurred image (or sharpen a noisy image). To produce different effects, all we need to change are the values in the filter matrix. The mathematical operations are exactly the same.

The ability to systematically extract edge information from an image can be useful for two reasons. First, as previously discussed, we could break down the edge information to construct a feature list for images we want to identify. Second, if we want to perform object tracking, all we really need is the outline of an object in a scene. The color or texture of the object adds extra information that will take time and resources to process, but color and texture information is not relevant to the location of the object in a scene. To perform object tracking, it's enough to know the outline of the object we want to track. For this reason, edge detection is often used as a preprocessing technique for various classical image-processing and computer-vision algorithms.

CONVOLUTIONAL NEURAL NETWORKS

Considering what we just learned about convolutions and their ability to extract information from images in a systematic way using a filter matrix, we might ask: What other filter matrices are there? And what other sorts of features can we extract from images in our quest to create computer-vision algorithms? In traditional image processing, the values inside the filter matrices were calculated following mathematical formulas and theorems devised by mathematicians and computer scientists. These had to be calculated with purpose; after all, they are extremely important, and the values inside the matrices determine whether the end effect is edge detection, blurring, or some other operation. Neural networks changed all of this.

The convolutional neural network is a neural network architecture based on a collection of filter matrices much like the ones we just discussed. But as you might have guessed, those filter matrices are not preprogrammed. Instead, in true AI fashion, the values of the filter matrices are discovered during the training process. A powerful CNN contains a robust collection of filter matrices that extract information useful in determining the content of an image.

CNNs are the classic neural network architecture for computer vision. As we saw in the previous chapter, CNNs are not the only class of neural networks that can be used for image analysis. Indeed,

MLPs are also capable of analyzing and classifying images into different categories of objects. The advantage of CNNs is that they greatly reduce the size of the model required to process images of increasing dimensions. The reason for this is that with MLPs, each pixel in the input image must have a connection to a neuron in the first layer, along with the accompanying weight for that connection. The larger the input image, the more connections required. But with CNNs, the weights for the neurons are shared across the entire receptive field of the neuron. That is, the same 3 × 3 filter matrix is reused many times to process all pixels in the image.

We begin by describing the CNN via a top-down approach. Initially, some concepts may be unclear and not fully developed, but we need to start somewhere. We base our CNN description on a classic architecture called VGG-16. VGG stands for Visual Geometry Group, the research group from Oxford University who first proposed the architecture. When it was introduced, it achieved state-of-the-art performance for image classification tasks on a number of popular research data sets, and it is still the backbone for many computer-vision tasks.

The VGG-16 architecture consists of thirteen convolutional layers followed by a standard three-layer MLP (a fully connected neural network similar to the ones described in chapter 1). The convolutional layers are a collection of filter matrices much like the one in our edge detection example. The task of the convolutional layers is to extract visual information, or cues, from the input image. The role of the MLP section is to classify the features extracted by the convolutional portion.

To understand how a CNN executes, let us first assume that our neural network has been trained, and it is ready to be used. That is, all the filter matrices in its layers have been programmed with useful values. To process an input image, we present the image (in this case, a 224 × 224 pixel image) to the first layer. In figure 2.3, we can see that convolutional layer 1 consists of sixty-four 3 × 3 matrices. For each 3 × 3 filter matrix, we process each pixel in the input image by the 3 × 3 filter, sliding the filter over each pixel across the input image, as described above. Since there are sixty-four such filters in this layer, the output of this first layer is sixty-four 224 × 224 pixel images. We call these images *feature maps*. The term comes from the fact that these images are the result of filtering operations aimed at extracting features from the original image. The process is the same as what we described earlier with edge detection or blurring filters; except in this case, instead of a single filter, we have sixty-four filters.

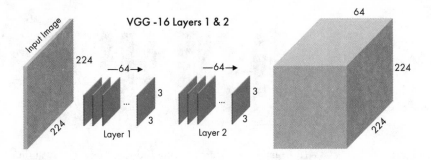

Figure 2.3 The VGG-16 neural network architecture consists of several convolutional layers, where each layer is a collection of 3 × 3 filter matrices. Layers 1 and 2 both contain collections of sixty-four 3 × 3 filter matrices. The output of these two layers is a volume of sixty-four images, called feature maps, the same size as the input image. *Author's rendering based on Karen Simonyan and Andrew Zisserman, "Very Deep Convolutional Networks for Large-Scale Image Recognition" (poster presented at the Third International Conference on Learning Representations, San Diego, CA, 2015), https://doi.org/10.48550/arXiv.1409.1556.*

The output of layer 1 is a feature map volume of sixty-four 224 × 224 pixel images. These images become the input to layer 2, which consists of another sixty-four 3 × 3 filter matrices. The input to layer 2 now gets processed by these matrices in a similar manner to layer 1.[11]

---

11. This is a simplified explanation of the process of performing a filtering operation over an input volume. The filters of CNNs have a channel dimension as well, and the channel dimension must match the channel dimension of the input volume. This means that layer 2, in fact, consists of sixty-four 3 × 3 × 64 filter matrices, where each image in the sixty-four-image input volume is processed by a different matrix in the 3 × 3 × 64 volume. Once the input volume is processed by a 3 × 3 × 64 filter, a single feature map is output.

The output of layer 2 is also a feature map volume of sixty-four 224 × 224 pixel images. These images will contain features that are different from those output by layer 1.

As we progress deeper into the VGG-16 architecture, layers 3 to 13 are similar to layers 1 and 2; what changes is the number of filters per layers. Layers 3 and 4 consist of 128 filter matrices. Layers 5, 6, and 7 consist of 256 matrices. And finally, layers 8 through 13 consist of 512 filter matrices (fig. 2.4).

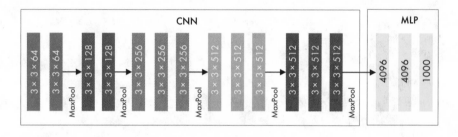

Figure 2.4 The VGG-16 neural network architecture is divided into a convolutional feature-extraction portion and an MLP classifier portion. The CNN section itself is divided into blocks of layers of equal numbers of filter matrices, with a MaxPool operation between blocks. The final output of this neural network as shown in the MLP section comprises 1,000 values. This neural network can classify objects into 1,000 different categories. *Author's rendering based on Karen Simonyan and Andrew Zisserman, "Very Deep Convolutional Networks for Large-Scale Image Recognition" (poster presented at the Third International Conference on Learning Representations, San Diego, CA, 2015), https://doi.org/10.48550/arXiv.1409.1556.*

As information flows through the different layers in the CNN, feature complexity builds. This is because the latter layers extract information from feature maps produced by the intervening layers. You can think of it as the deeper layers using the features extracted by earlier layers and putting those features together to build higher-level concepts. Specifically, the difference in complexity means that the early layers typically learn to extract edge and color information, whereas the deeper layers learn to detect texture and such conceptual information as eyes or whole faces or even motion.

Between blocks of convolutional layers, an operation takes place that's very common to CNNs—the MaxPool operation (fig. 2.4). For each 2 × 2 pixel window in each feature map, the highest value is chosen, and the rest are discarded (see fig. 2.5 for an explanation). This has the immediate effect of reducing the size of the images by half for subsequent processing, which has the benefit of speeding up the computations in the latter layers. Reducing the size of the images by half means that the feature map sizes after the first MaxPool operation go from 224 × 224 to 112 × 112 pixels. The second MaxPool operation reduces it to 56 × 56 pixels. By the time we get to the end of the CNN portion of the network, the size of the feature map volume is 7 × 7 × 512 pixels—that is, 512 images of 7 × 7 pixels each.

12 Pixels

1	3	10	0	5	98	12	1	55	101	1	9
18	88	43	23	1	4	5	254	0	12	135	56
6	2	55	65	80	34	100	5	3	78	0	1
27	12	0	10	19	11	0	2	15	102	4	56
60	100	2	98	113	2	20	91	26	2	34	123
2	1	14	0	2	13	89	11	200	33	93	0
15	65	22	24	45	5	65	4	65	5	23	34
1	45	62	13	2	200	22	15	11	1	120	65
0	250	120	44	4	1	0	41	15	5	11	3
31	16	100	1	39	45	101	54	3	27	0	79
16	40	3	111	34	3	11	79	76	1	5	82
1	1	34	23	0	100	2	5	0	88	2	180

12 Pixels

MaxPool

6 Pixels

88	43	98	254	101	135
27	65	80	100	102	56
100	98	113	91	200	123
65	62	200	65	65	120
250	120	45	101	27	79
40	111	100	79	88	180

6 Pixels

Figure 2.5 The MaxPool operation serves to reduce the size of the data that must be processed in each block. It is performed by selecting a 2 × 2 pixel window on the top left of the input image and then selecting the pixel with the highest value in that window. This pixel is kept and becomes the first pixel in the output image; the other pixels are discarded. We then slide the 2 × 2 pixel window to the right and continue processing the input image. In this example, our first 2 × 2 pixel window contains the following pixel values: 1, 3, 18, 88. The highest value is 88, so we keep this pixel and discard the rest (*right*). The next window contains: 10, 0, 43, 23. The highest value is 43, so we keep this pixel and discard the rest. Thus, we reduce the input image from 12 × 12 pixels to 6 × 6 pixels.

The intuition behind max pooling derives from interpreting the values in each 2 × 2 region of the feature maps as indications of important features in the data. By selecting the largest value and preserving it for subsequent processing, we retain the most important feature in the 2 × 2 window. As with many subjects related to artificial neural networks, however, the intuition behind certain operations is sometimes added as an ad hoc explanation after a concept has been developed. In reality, much of what we do with neural networks is done because it was empirically found to work. In the case of MaxPool, the main benefit (and what has influenced the implementation of this operation) is reducing the image size so that subsequent layers have less data to process. Once a concept like this is implemented, it is tested against well-known data sets and different neural network architectures. If the concept works, it is kept, and other people incorporate it into their research. This means that we often don't end up with a theoretical proof of why or how it works. If this sounds like a less-than-rigorous approach to science, in some respects it might be. The benefit of this approach is that the field of artificial intelligence has experienced an unprecedented explosion in results and solutions in recent years. Whether this is a good thing or a bad thing depends on what our ultimate goals are. If the goals are to advance the AI field and solve discrete problems like image classification or natural language processing, clearly this approach is working. If the goal is to fundamentally understand "intelligence" and build a model of the human brain, I suspect we will need a more rigorous approach where we fundamentally understand the gains and losses of each operation: For example, why does MaxPool really work, and what is the cost of the values that we discard in a more generalized sense?

The last layer in the CNN just before the classifier section outputs 512 7 × 7 pixel feature maps. Before proceeding to the MLP section, we take each 7 × 7 pixel image (or feature map), unroll it into a single *flat* vector of 49 values, and attach each of the 512 vectors to the end of the previous one to generate a 25,088-value vector (7 × 7 × 512 = 25,088). When we present this vector of 25,088 values to the classifier, this is the data it uses to *predict* the class of the object in the original image.

Note: we omitted a slight complication in the operations of the CNN. For each filter neuron, at each processing step where we filter a 3 × 3 input volume, the result of the filtering operation is further processed by a nonlinear function f. After all, a CNN is still a neural network. Each 3 × 3 filter matrix is itself a neuron, and as we know, neurons require an activation function to decide when to fire. In the case of the VGG-16 architecture, the nonlinear function f used is a ReLu function. Recall that this function takes an input value, and if the value is negative, the function outputs 0; if the value is positive, it returns the same input value. Suppose that the result of our filtering operation is −4; then, $f(-4) = 0$. That is, the value produced by the neuron will be 0. If the result of our filtering operation is 4, $f(4) = 4$. The value produced by the neuron will be 4.

The filtering process in the CNN extracts and combines information from the feature maps in a manner similar to the edge detection example and creates a new set of images (i.e., feature maps) consisting of a specific set of features. These may be edge information or texture information, and the deeper we get into the neural network, the more and more complex the extracted features become. If the neural network was trained to detect animals or human faces, in the deeper layers, we might find neurons that are sensitive to higher-level concepts like eyes, ears, or whole faces. If we think of the filtering process as similar to extracting some elements out of a bucket of sand using a series of sifters, we can imagine that each subsequent sifter lets finer-grained material through. In the end, what we are left with is a distilled version of the bucket of sand that we started with. Thus, we can say that the final output of the convolutional section is a distilled version of the original image. If the neural network was trained correctly and is working, the output consists of a compressed, smaller, and more potent version of the original input image.

Once we produce that distilled version of the original image (in our case, the original image has 50,176 pieces of information, or 224 × 224 pixels), this distilled version is sometimes called a *latent representation*, or *latent vector*. In our example, the latent representation has 25,088 pieces of information. This represents a reduction by 50 percent in the amount of data that will be used to classify the object in the image. Remember, the CNN section does not perform classification. It simply extracts information from the image in the form of feature maps and combines this information into a latent, or distilled, version of the input image. The latent representation then gets processed by the classifying section, and it is this section that makes a prediction as to what class of objects is hiding in that latent representation.

You may be wondering about the intuition that the latent representation contains all the information we need to make a prediction on the object we are trying to detect. After all, how do we know that by losing 50 percent of the original input we haven't lost some vital part of the object? We began this chapter with the example of trying to detect a cat in an image. In one of the images we discussed, our cat was lying on a bed. If we think of that image as a collection of pixels and each pixel as a unique piece of information, how many of those pixels are vital to our understanding that there is a cat in that picture? Put another way, if we opened that image in Photoshop and deleted or changed a few pixels, would we no longer be able to detect the cat in the picture? It turns out that if our goal is to predict "cat," most of the information in that image is superfluous. If we removed the bed and all the background from the image, we would still be able to detect the cat. In fact, if we removed all the pixels that describe the fur and texture of the cat and left just the edge information, we would still be able to detect the cat in the image.

Hopefully, we can now start to see why compressing the original image before classification might not be such a bad idea. For any given image, there are always pixels that could be removed without losing vital information about the subject. The goal of training our CNN is to have the algorithm *learn* which information is OK to remove and which isn't. This in part depends on what our goal is in our use case. For example, if we want to detect cats in images but we are not worried about which breed of cats, all we really need our algorithm to do is encode information about the shape of cats; texture information isn't as important. But if we decide that we want our algorithm to detect not just the shape of the cat but also whether it is a Siamese or a Bengal cat, then encoding texture information in the latent vector might be a good idea. This makes sense because different breeds of cats also tend to have different colored and textured furs. When we are discussing AI concepts, we need to always keep our goals in mind. In other words, what is it that we are asking the algorithm to do? As we will see, when we train our algorithms, they are trained to meet the goal that we set out for them. So if our goal is not well defined, the algorithm will not meet our asks, just as if we train a CNN to detect the general shape of cats, we cannot expect it to also detect breed.

USE-CASE EXAMPLE

Let's see how we might employ our CNN. Suppose that we are tasked with creating a computer-vision system for a sorting facility. We want a machine to sort a list of artifacts into different boxes. To do that, the machine needs to distinguish between different classes of objects. In our example, the machine needs to sort through different types of batteries: 9V, AAA, AA, and button batteries. So we have four classes of objects that our system needs to recognize. This means that our VGG-16 neural network will have four different output channels, one for each class. Note that in figure 2.4, there are one thousand different outputs. This is because the classic VGG-16 neural network was trained to recognize one thousand different classes of objects using the popular ImageNet data set, which contains over 14 million images. The output layer can be modified, however, to output as many channels as there are categories in the problem we are trying to solve.

So how does our automated sorting machine work? First, it uses a camera to take a picture of the object in front of it. The image is then presented to the neural network, which processes it and outputs a set of probabilities spread across the four output classes. Table 2.1 shows that the neural network has output classes 0, 1, 2, and 3, one for each class of object it needs to identify.

Table 2.1 Four Classes of Batteries Our Neural Network Must Identify

Class label	Class
0	9V
1	AAA
2	AA
3	Button

When the assembly line starts running and brings a set of batteries for our sorting robot to distribute among different battery bins, the camera snaps a picture. The first item to be sorted is a 9V battery. The image is presented to the VGG-16 neural network, which outputs a prediction for which type of battery it believes it is. On a well-trained neural network, the outputs might look the ones in table 2.2.

Table 2.2 Probability Outputs for Each Predicted Class

Predicted class	Prediction
0	0.98
1	0.01
2	0.007
3	0.003

In table 2.2, the prediction represents the neural network's confidence, as a probability, that the item in the image belongs to the class represented by the output class. So here we see that the neural network calculated a 98 percent probability that the image is a 9V battery, a 1 percent probability that it's a AAA battery, and so on. Note that since we are discussing probabilities, the sum of the output values must equal 1 (100%). We have just seen how CNNs communicate with us. It's really very simple. They output a set of probabilities through their output channels, and we pick the highest probability as the network's prediction. In practice, the difference between the predictions across the output channels isn't always this stark. For example, it is possible that for some images of AAA batteries, at certain viewing angles, the predicted difference between a AAA and AA classification is much closer, like in table 2.3.

Table 2.3 Probability Outputs Conveying a Neural Network's Lower Confidence Level on the Correct Classification

Predicted class	Prediction
0	0.025
1	0.65
2	0.32
3	0.005

In table 2.3, we see a 65 percent probability that the battery is a AAA battery and a 32 percent probability that it's a AA battery. We can understand some hesitation on the neural network's part in distinguishing between AA and AAA batteries since they are quite similar, with overall size being the main differentiator. What would be surprising is if the neural network were to analyze a 9V battery and output close predictions for 9V and button batteries. You would intuitively expect that 9V batteries and buttons batteries have features that are distinct enough for their predictions to be far apart.

Whether we are classifying images or numerical data samples, as we did in the previous chapter, the concept remains the same: Neural networks have one output channel per class (if we need to classify ten different items, the network will have ten different output channels). The network outputs a probability that the item we are trying to classify belongs in each possible category. We then choose the output channel with the highest probability and say that the image belongs in the class that is represented by that output channel. In principle, this is no different than the simple MLP we discussed in the last chapter.

TRAINING AND ARCHITECTURE DESIGN

At this stage, you should have a good understanding of what the convolutional section of a CNN does (extract a set of important features from an image) and how it does it. It is just a series of operations, mostly multiplications and additions (with a nonlinear function processing the result of each weighted sum), which have the effect of extracting specific elements from the input image. These in turn are key to detecting the subject of the image. That is exactly why this process is called *filtering*: just think of how adding different filters to a camera lens reveals different aspects of our world, which might be harder to see with the naked eye.

There are still two important pieces of information we have not discussed. The first one is how we decide on the architecture of the CNN. For example, why does VGG-16 have thirteen convolutional layers, and why does the third block have layers of 256 neurons while the fourth and fifth blocks have 512-neuron layers? How do we know we have enough convolutional blocks? How do we know we're not missing a few blocks? The second piece concerns how we program each 3 × 3 filter in each layer. How do we select the numbers that should go in each cell of the 3 × 3 filters?

Unfortunately, deciding on a CNN's architecture is not a simple process. There aren't strict rules for how we choose the architecture of a neural network to solve a given problem. There are general considerations that can guide the intuition for the number and size of each layer. The larger the model, the more trainable parameters it has (i.e., the more 3 × 3 filters that must be programmed); the more trainable parameters it has, the larger our training data set needs to be. Why are the sizes of the data set and the model related? If we have a small data set but a neural network with many millions of parameters, we might find that our neural network "memorizes" the training data. That is, the neural network will follow the training data too closely. This is bad because a small training data set may not be a great example of the real world, so we are not really preparing the neural network for succeeding in the real world if we allow it to "memorize" all the data. Similarly, if we have a large training data set but a very small neural network, we may not have enough trainable parameters to adjust for extracting an appropriate diversity of features from our data set.

A simple rule of thumb is to not make the model too large if our training data set isn't large enough. In practice, however, researchers fine-tune the shape and architecture of the network simply by trying different things and seeing what works. So we might add or remove layers until we have a solution that works. Over the last decade, researchers have discovered neural network architectures that are better suited for particular classes of problems, from computer vision to natural language processing to outplaying a chess grand master. Once a base architecture is found to work well for a specific class of problems, other researchers typically iterate over established model architectures and make little modifications here and there until the model works for their particular use case. The important thing to understand is that there are no commandments that describe how to design a neural network to solve a given problem. Instead, we start from an established architecture known to work for our class of problems (e.g., classification of human faces or classification of chest X-rays) and then slowly add/ remove layers and blocks until the model starts to perform well.

Now we can address the second issue. How do we program the values in each cell of the 3 × 3 filter matrices? If we remember from the edge detection example at the beginning of the chapter, the values we put into the filter matrix are very important. If we put in arbitrary values, we will get a resulting image that will not contain any useful features. It might look corrupted or like random noise. If we put a special set of values known to filter edge information, we get a resulting image that contains just the edges of the original image (fig. 2.1). It should be clear by now that these values do the very important job of extracting information from the input image. Our goal is to create an algorithm that can extract the right features from an input image that can lead to the classification of the subject. This is precisely what the training process does.

During the neural network's training process, our aim is to *discover* the values that must be put in each 3 × 3 filter so that the right set of features are extracted from our images, which, when presented to the classification section of our network, can lead to the correct prediction of the object's class. The reason CNNs end up using so many filter neurons is that vision requires a complex system where features are extracted at different levels and combined to create higher-level concepts in the latter layers. The impressive achievement in the training process is that we establish a set of steps that systematically discover a set of filter values that leads to the algorithm predicting the correct class to remarkable levels of accuracy. All of this without explicitly telling the algorithm what's special about cats or humans or AA batteries!

To train a neural network, we begin with a training data set— which, for a computer-vision algorithm, is a specially crafted collection of images and labels. The labels represent what we call the *ground truth* for the class of the image. If we want to train a model to differentiate between humans, cats, dogs, and bears, our data set must consist of many images of humans, cats, dogs, and bears. Each image in the data set will contain a label that tells our neural network the class of the image. As we saw earlier, the label is just a number that we assign to a particular class. We could define our training data set to have label 0 for humans, 1 for cats, 2 for dogs, and 3 for bears. Then every image of a dog in our data set will have a label of 2.

Initially, the parameters in the neural network—that is, all the values in the 3 × 3 cells of the filter neurons—are initialized with random values. Since the values are randomly selected, the neural network isn't very good at extracting useful features from the input images, and the predictions are highly inaccurate. But as training progresses, the values in the filter neurons get adjusted, similar to how one might slowly turn the dial on a radio to tune to a specific frequency. As the filter values get adjusted, the neural network gets better at extracting useful features in each layer, and the predictions get more accurate.

To understand exactly how the network parameters are adjusted through training, we need to understand what happens when the network makes a mistake. To make this example easier to follow, let's scale the problem down to a binary classification task. In honor of the great sitcom *Silicon Valley*, let's train our network to recognize food items as either "hot dog" or "not a hot dog." In this example, the "hot dog" class will have the label 1, and the "not a hot dog" class will be labeled 0. Since we are just beginning to train our neural network, our parameters all consist of random values. Now we go and present an image of a pizza slice to our network. Since the network hasn't been trained, the probability that it will correctly predict "not a hot dog" is 50 percent, so let's go ahead and assume that it makes a mistake and predicts "hot dog." Remember that neural network predictions are done using numbers, and in the binary classification example, the output will be a value between 0 and 1. An output of 1 represents the neural network being 100 percent confident that the image is a hot dog. An output of 0 represents the neural network being 0 percent confident that the image is a hot dog (another way to say this is that the neural network is 100 percent confident that the image is not a hot dog). Any output between 0 and 1 signifies a confidence level for the likelihood that the image is a hot dog. Because the mistake we are discussing happens during the training phase, the image of the pizza slice we are presenting to the network has a label that represents the image's true category.

In our example, the neural network has made a mistake and predicts a value close to 1 (0.89). This prediction tells us that the neural network is 89 percent confident that the image it saw is that of a hot dog. The training algorithm happens to know that this image is in fact not a hot dog because it knows the true label of the image, which happens to be 0 ("not a hot dog"). Now the training algorithm needs a method of determining how far its prediction was from the true class of the image. We call this method a *loss function*. The loss function provides a way to measure how far the model's predictions are from the ground truth. By knowing how far off the predictions are, we can calculate how much we need to change the model's parameters so that next time the predictions will be a bit closer to the ground truth. Intuitively this makes sense: if the predictions are very far from the ground truth (as in our example, where 0.89 is quite far from 0), we need to modify our model's parameters by a larger amount than if the prediction had been closer to the ground truth (say, 0.2).

In chapter 3, we examine a more detailed example of how the parameters of the model are updated; for now, we are just trying to build intuitions. We want to understand how in principle a generic algorithm can modify parameters of a neural network to progressively make better predictions about the data it is analyzing. What we have just learned is that the training algorithm uses a set of training images, presents the images to the model, gets predictions from the model, measures how far the predictions are from the ground truth (which we know because a training data set contains labels for all the samples), and uses that information as a measure for how much the model's parameters should be updated. Then, the next time the model is presented with that same image, the prediction will be closer to the ground truth.

The process for updating the network parameters is called *gradient descent through back propagation*—a very elegant algorithm that methodically calculates an amount by which to change each network connection (or each cell in the 3 × 3 filter neuron). The network connections have values that range between 0 and 1; back propagation uses calculus (Remember calculus? It turns out it's useful after all!) to calculate how much a change in each connection contributes to the error in the prediction and adjusts each value in the direction that minimizes the error. We address this in greater depth in chapter 3.

An intuitive way of picturing the process of tuning the neural network parameters to produce accurate results is to think of a neural network as a mathematical function that transforms an input value into an output value. A mathematical function such as $f(x) = x + y$ transforms an input x by adding y. Suppose we used an input of 4 and randomly initialized y to 1. The output of the function would be 5. Now suppose we want to get the function to output a value of 9. We need to figure out how to modify y to get our desired output. With y initialized to 1, our output is 5, but the desired value is 9. We can use a simple loss function (introduced in our "hot dog" / "not a hot dog" example) and subtract the desired output from the actual output: $9 - 5 = 4$. This tells us that our output is 4 points off the mark. We can now set y to 4 and try again. This time we get $5 + 4 = 9$, which matches our desired output. In principle, this is exactly what we are trying to achieve with the neural network. It truly is a mathematical function in very much the same way. The difference is that instead of a single parameter y, there are millions of parameters, so the process of updating those connections is more complex, and the updates must be made more slowly, inching closer to the answer over time.

LEARNING TO MODEL A DISTRIBUTION OF FEATURES

In our mathematical function example, there is just one right answer: 9. So we can update parameter y at once to a setting that gives us the answer we want. We can describe $f(x)$ as mapping an input to an output value. For example, by setting the y parameter to 4, we have mapped input 5 to output 9. In our image classification example, the "right answer" is more complicated. We want the network to recognize images of hot dogs, but there isn't just one single image of a hot dog. Think about how many ways we can make an image of a hot dog look (*yum*). And the network needs to be able to recognize them all. Put another way: Think of an image as an arrangement of pixels. In how many ways can we arrange the pixels so that it results in something resembling a hot dog? Considering that the resolution of the image is fixed, the answer is not infinite, but it is clearly quite large. Therefore, when we update the connections of our network, we can't update them in such a way that it can recognize only this one image of a hot dog. It needs to recognize any image of a hot dog.

We dive deeper into the mathematics behind the classification when we discuss linear and logistic regression. For now, we just need to get a feeling for what is happening behind the scenes. In our simple function $f(x)$, when we turn the dial on parameter y and adjust its value, we are trying to home in on a specific output. With AI algorithms like neural networks, what we are trying to do is adjust the many dials (millions of dials) of the function, but instead of trying to map a specific input to an output, we are trying to map a range of inputs to a specific output. We can think of every class of objects we are trying to classify as consisting of samples that share common elements. These are the attributes that determine whether they belong to that class. For example, we might have ten images of different-looking hot dogs, all of which are still recognizable as hot dogs—just like we can have ten images of different cats that share enough commonality to still be recognized as belonging to the class "cats."

Consider this: two different images of a hot dog may look very different from each other, but we still know they are not so different as to be confused with a cat. Remember that images are arrays of pixels, and pixels can be interpreted as numerical values. This allows us to consider an image as a vector in some multidimensional space where all possible images of hot dogs point in some similar direction. Similarly, all images of cats would consist of vectors that point in roughly the same direction. These vector groupings we call *distributions*, because the samples are arranged, or distributed, close to each other in some hyperspace. When we assign a class to a group of samples, we are determining that they belong in the same feature distribution. The role of an AI algorithm is to learn the shape and location of the distribution in the multidimensional vector space where all things exist.

All right, now we have gleaned enough information to gain access to the secret behind the madness. The power of neural networks is that they learn to model the distribution of features for the classes they are trained to classify. By learning the shape of the feature distribution for each class of samples, they learn to map a range of possible inputs to each output. This is what makes it possible for a neural network to function beyond training. That is, once the neural network is trained, it can classify brand-new images it never saw during training. Neural networks are not matching algorithms that store a database of known images and later reference them. Neural networks learn a space of possibilities for the features that define an object's class. Later, when presented with a brand-new sample, they can check which of the learned distributions more closely resemble the features in the new sample and thus classify it.

It is important to note that the very quality that makes neural networks powerful can be a source of weakness. You see, learning the distribution of features for a class of samples is a powerful technique because it means that for unseen data, the network only has to check whether the new sample exists in the same distribution space. This is a fancy way of asking, when we interpret the new input image as a vector, whether that vector is pointing in roughly the same direction as one of the classes of objects the network is trained to identify. The downside is that the network has learned the shape of the distribution based on the training data. If the training data set is not large enough, or if it is biased toward some population sample, then the learned distribution will not match the real world, and the model will perform poorly outside the lab environment.

For example, suppose the neural network is trained to identify cats using only images of Siamese cats, and we later ask it to identify an image of a white Persian cat. These two cat breeds are distinct enough that the neural network may not find that they belong to the same distribution. To train a neural network to be robust enough to recognize any cat image presented to it, we must train it against a data set consisting of thousands of images of all cat breeds. The less robust the training data set is, the more limited the learned distribution will be, and therefore, the more mistakes we can expect in a production environment.

Suppose that the neural network performs quite well at recognizing cats in general. What if we wanted to train it to recognize different breeds of cats? The training process remains the same, but the data set would need to change. This is a beautiful quality of neural networks. To train the neural network to recognize different classes of objects, we simply need to change the training data set. The model and the training algorithm remain largely unchanged. To teach the network to recognize different breeds of cats, we would require a data set that contains thousands of images of the different breeds of cats we want to identify. The neural network can then learn the distribution of features of each cat breed. But beware! If any one class of cats is underrepresented in the training data set, the learned distribution will be a poor match for the real distribution of features for that class, and the neural network will not accurately represent that class.

ARTIFICIAL VS. BIOLOGICAL VISION

So far we have discussed CNNs, which are a class of artificial neural networks, as mathematical functions mapping a range of inputs to specific outputs as they learn feature distributions. To some, this view might reduce the romantic charm of artificial neural networks in much the same way that learning a magician's trick tarnishes the performance. This might be true if we expected artificial neural networks to resemble the mushy neurons in our own brains capable of complex thought and consciousness. To some extent, however, this disappointment is the result of undue expectations on our part. Luckily, unlike the magician's trick, which is simply an illusion, artificial neural networks are real algorithms drawing from concepts across multiple disciplines spanning decades of research, and they work! Also, as it turns out, the more we learn about the mushy stuff in our noggins, the more we see that the ethereal and mystical process of biological vision is just a series of computational steps.

Now that we have a basic understanding of computer vision, it's time to spend a few moments going over biological vision systems and seeing where art imitates reality. This is what we have all been waiting for, isn't it? We want to see how well our algorithms imitate us. One of the most fascinating aspects of artificial neural networks like CNNs is when these complex systems exhibit emergent properties that resemble biological systems. Emergent properties are properties that arise from a system without being purposefully built into the system. But before we can dive into the emergent properties of CNNs, we need to discuss some of the architectural or physiological parallels between biological and artificial neural networks. This chapter has been all about vision, so we will continue to follow the vision thread as we explore parallels between artificial and biological systems. We have learned that artificial neural networks are built as stacks of layers made up of neurons, where connections and information flow from layer to layer. It turns out that, at least in principle, our vision system is quite similar.

Let's start with an overview of the most important elements of our vision system. For the purposes of this book, we don't necessarily need to understand each component of our vision system. We will do the best we can, however, to start with a robust description of the components involved; this way, we can at least start to see how vision involves many different processes and structures working together, beginning with simple concepts and building complexity. The vision system of mammals consists of three main components: the retina, the lateral geniculate nucleus (LGN), and the visual cortex in the brain. The visual cortex itself is divided into five areas: the primary visual cortex (V1) and the secondary visual cortex (V2), as well as V3, V4, and V5.

Information processing for vision begins in the eye itself, at the retina. The retina consists of several layers of cells, beginning with ganglion cells and followed by bipolar and horizontal cells connected to a photoreceptor layer consisting of rod and cone cells. Rod cells are typically concentrated around the outer edge of the retina and are mostly responsible for night vision. Interestingly, our eyes have different structures for processing colors and grayscales. Rods seem to play almost no role in color vision, which might explain why it is very difficult to distinguish between different colors in dim environments. They also take significantly longer to adapt to light compared to the color-sensitive cone cells. This is why it takes a long time for our eyes to adjust when we move from a well-lit room into a dark room.

For color vision, we have three types of cone cells, each sensitive to a different frequency of light: red, green, and blue. You might know of someone who is color-blind. People who are color-blind have sustained damage to one or more types of cone cells. Depending on the types of cone cells that are damaged or not functioning, the individual will be unable to perceive certain color ranges. The most typical form of color blindness is a deficiency in perceiving color in the red and green frequency range. A more severe but less common form of color blindness is a deficiency in perceiving blue and yellow color ranges. As we will see, these color pairings—red-green and blue-yellow—are also associated inside artificial neural networks.

The cone and rod cells are connected to bipolar cells, which in turn are connected to ganglion cells. The ganglion cells are a very important layer of cells that connect the retina to the LGN and the primary visual cortex in the brain. An interesting property of the retina and the organization of these layers is that the ganglion and bipolar cells are directly in the path of the light as it travels into the eye toward the photoreceptors. You would think that one would place the photoreceptors in the outermost layer, free of interferences; evolution had other plans. Thankfully, the layers of cells in front of the photoreceptors are mostly transparent and offer little disturbance to the light.

At a high level, the photoreceptor layer detects photons that enter the eye. The detection process generates signals between the photoreceptors and the ganglion cells. Here we encounter the first information-processing problem in our vision system. There are approximately 100 million photoreceptors in each eye but only about half a million ganglion cells. So as information proceeds from the eyes to the LGN and further toward the primary visual cortex, a significant amount of data compression takes place. That is, we start with roughly 100 million data points in each eye and reduce that to roughly half a million data points by the time information leaves the eye en route to the brain. Does this sound familiar? Data compression is a quality of information processing that is not unique to biological vision systems; indeed, data is compressed and expanded as it travels between different stages in the visual system, and as we saw, the same is true with artificial neural networks.

In the 1950s, Stephen Kuffler conducted a series of experiments on anesthetized cats that showed information processing starting even at the level of the ganglion cells in the retina. He and his team placed an anesthetized cat on a table with its head facing a screen and the eyes held open. They then showed an image of a black background with a white spot moving against the black background. As they moved the white spot around the screen, they measured the output signal on the axons of specific ganglion cells connecting the retina to the LGN. They noticed that some cells were very sensitive to what they called "on center, off surround" areas, while other cells were sensitive to "off center, on surround" areas. This meant that when the spot of light was at the center of the cell's receptive field, the "on center" cell fired a long burst of signals. As the spot moved off the receptive field, the cell moved from steady firing to not firing.

They also changed the white spot image to a white spot with a dark center, like a donut shape. As they moved the donut shape around the screen, they noticed that when the shape entered the receptive field of some ganglion cells, the cells fired when the dark spot was at the center of the receptive field. As the dark center moved off the receptive field, the cells stopped firing. Furthermore, for both the on-center and off-center cells, they tested what would happen if the center of the spot increased in size to encompass most of the cell's receptive field. That is, for the on-center/off-surround cells, they increased the size of the bright spot, causing the receptive field of the cell to only see the "on" portion of the image as the off-surround area moved outward off the cell's receptive field. In this case, the cell stopped the steady firing. When the researchers reduced the size of the bright spot so that the cell perceived the on-center and off-surround area again, the cell went back to a steady signal. They encountered the same results for off-center and on-surround cells. As the dark spot increased to lose the on-surround area, the cell stopped firing, and as the dark spot's size was reduced and the on-surround area came back into the cell's receptive field, the firing commenced again.

A decade later in the 1960s, David Hubel and Torsten Wiesel extended these experiments to show that on-center and off-center information perceived by individual ganglion cells in the retina was combined in the LGN and V1 area of the visual cortex to construct bars, or edges, of on-center/off-surround and off-center/on-surround sensitivity. Do you realize what this means? This shows how complexity builds as information moves from layer to layer from the retina to the brain (fig. 2.6). First, the eye perceives discrete points of white on dark background or dark on white background. Later, these individual data points get combined to build more complex information, such as edges. Furthermore, the Hubel and Wiesel experiments showed that special types of cells in the V1 area of the visual cortex called *complex cells* used the information gathered by ganglion cells at the retina to construct edges of different orientations. That is, some complex cells are sensitive to horizontal edges while other cells are sensitive to diagonal edges. We have already discussed the importance of edge detection in the context of computer vision. We now see that vision in biological systems, or at least in mammalian organisms, also starts with edge recognition. For their work on visual processing in biological visual systems, Hubel and Wiesel received the 1981 Nobel Prize for Physiology or Medicine.

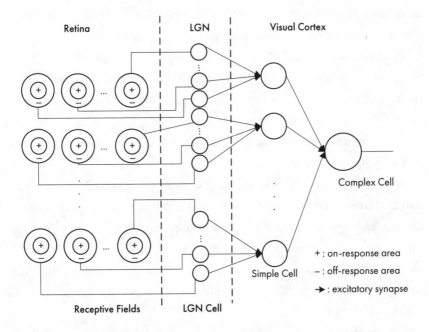

Figure 2.6 Hubel and Wiesel's hierarchy model. Information flows from the retina to the brain's visual cortex in a cat. It's hard not to see a strong resemblance to artificial neural networks. *Author's rendering based on Hubel and Wiesel 1962.*

The primary visual cortex (V1) is by far the most studied region of the brain's visual cortex. This area is divided into six different layers, 1 through 6. The estimated number of neurons in the primary visual cortex of an adult human is around 140 million. The primary visual cortex is the main receptor of information from the LGN and retina. It is great at pattern recognition. As information proceeds into the latter regions of the visual cortex (V2–V5), higher-order, more complex information is captured. Although these regions have not been studied as much and are not as well understood as the primary cortex, it is understood that such global concepts as faces and textures are recognized in these latter regions. These are called higher-order concepts because they are more complex than pattern or edge recognition. After all, to recognize a whole face, we need to combine many more primitive concepts like edges, color, and even depth. This is an important discovery, so it bears repeating. There are complex cells in the visual cortex that are sensitive to complex structures, such as faces, in their receptive fields. This means that if we measure the output of these cells, we notice a spike in their signal when a human face crosses their receptive field. This is surprising because, intuitively, we would expect that a single neuron would be too simple a device to detect a complex feature like a whole face; it shows the hubris in our intuitions and why sometimes to make progress, we must abandon all that makes sense. Indirectly, we also know that complex structures like faces are handled by special components in the brain from studying a condition known as *prosopagnosia*. Prosopagnosia typically affects people from birth, and it is a condition whereby the afflicted individual is unable to recognize faces but has otherwise perfect vision. These studies have found evidence that faces are detected by very specialized cells or regions of the brain.

Now that we have a basic understanding of how vision works in both computer and biological systems, we can spend a few minutes discussing some of the parallels between biological and artificial vision, as well as what we have been referring to as emergent properties of neural networks. In figure 2.6, we saw Hubel and Wiesel's hierarchy model of the visual system of a cat. Nearly all image-processing operations we have discussed thus far, including most artificial neural network architectures, involve weighted sum operations—that is, an operation where the values of an input signal to a neuron (such as the pixel values in an image) are multiplied by the weights of the neuron's input connections, and the results are summed. Hubel and Wiesel proposed that weighted sum operations are also present in the neurons of the cat's (and by extension our) visual cortex.

Experimental studies involving complex cells later found evidence that suggests that weighted sum operations do indeed happen in some neuron cells in the brain. As we have already seen, both biological and artificial neural networks are composed of layers of neurons with connections between the layers and information flowing from layer to layer. Vision in the mammalian visual system starts at the eye, with the first few layers of neurons, the ganglion cells, detecting simple on/off regions. These signals are processed by the primary visual cortex to form edges and perform pattern recognition.

We saw with the CNN that a similar phenomenon arises in artificial neural networks. The first few layers are sensitive to edges and simple patterns. As we move deeper into the CNN, we find that the latter layers contain neurons that respond to more complex information. We find neurons that are responsive to whole faces, to eyes, and to texture information, just like we saw with biological systems! What is really fascinating is that CNNs were not predesigned to perform edge recognition in the first layers and detect higher-order information, like whole faces, in the latter layers. This is an emergent property of neural networks. In fact, we get the same distribution of information (the first layers responsive to edges and latter layers building complexity) for any neural network architecture we build, from CNNs to MLPs. So how is it possible for artificial systems to break down visual constructs in a similar manner to biological systems without being explicitly designed to do so? The definitive answer to this question is not really known.

Before we proceed, take a moment to digest what we have just said. It is quite possible that thus far in your life, you have never encountered an example of a system built by humans that performs a set of tasks that were not designed a priori. When I started researching and learning about neural networks, I found that the stuff we don't know about these systems (the stuff whose presence itself is especially remarkable because we build these systems) is far more interesting than the things we do know. So let's go back to our question: How is it possible that these emergent properties in artificial systems arise to resemble biological ones? As I said, we do not know for sure, but an explanation that I like is as follows: we build complex systems from simple rules, and as the system grows in size, the growth in complexity will generate behavior we could not have predicted a priori.

Take the Game of Life as an example. The Game of Life, created by the British mathematician John Horton Conway in 1970, is not really a game where players compete to reach a certain goal; instead, it is a simulation of a world that is bound by only four rules. The world starts with a grid of cells and proceeds based on the following rules:

1. Any live cell with fewer than two live neighbors dies of underpopulation.

2. Any live cell with two or three live neighbors lives on to the next generation.

3. Any live cell with more than three live neighbors dies of overpopulation.

4. Any dead cell with exactly three live neighbors becomes a live cell.

Life begins by selecting a few cells in the grid and setting them to "alive" and then letting the simulation run for a number of generations where those four rules are applied for each generation. Following just four simple rules, incredibly complex patterns are generated. The most common pattern is called "the glider," which consists of a group of cells that move around the grid. There is a whole Wikipedia page dedicated to interesting Game of Life patterns that people have found. Interestingly, new patterns are still being found by Game of Life enthusiasts.

So what does the Game of Life have to do with vision? When it comes to computer vision and artificial neural networks, the important part is that the system is built from simple rules. That is, if we want to detect elephants in an image, we do not set out to break down concepts of what makes elephants unique (long trunk, big ears) and build an algorithm around detecting those. Instead, we do not make any assumptions as to how we are going to detect elephants. We start from an array of discrete points of light (pixels) and build an algorithm to learn to extract and combine the features that will determine whether there is indeed an elephant in the image. It might be that the best way for a system to detect visual information from first principles is to start by detecting edges and then putting those edges together to build more complex features. In fact, this might be an unavoidable consequence of building knowledge from a set of pixels (or photon detection in our eyes), considering that regardless of the neural network architecture we design, we still get edge and pattern recognition at the first layers. This makes it clear that the distribution of features among the layers is not dictated by the network architecture itself but by some fundamental truth about vision systems. Amazingly, our artificial neural networks learn this through their training process by minimizing a loss function, and biological visual systems converged on a similar approach through evolution.

Feature distribution is not the only emergent property of neural networks resembling biological systems. When we look at the features themselves, we notice other interesting similarities. We already know that the first layers of a neural network encode edge information. That is, the neurons in the first layers of the networks are sensitive to edges in different orientations. Some neurons (remember: in a CNN, a neuron is a 3×3 filter) are sensitive to light-on-dark edges. Other neurons are sensitive to edges defined by color pairings (e.g., red on green and blue on yellow). What is quite interesting about these color pairings is that red/green and blue/yellow are closely associated in the mammalian visual system as well. As we previously discussed, people who suffer from color blindness have trouble seeing colors in the red/green or blue/yellow range, with red/green deficiency being the most common form of color blindness. Yet here we have artificial neural networks discovering that it is important for vision systems to encode contrast information between red, green, yellow, and blue light as well. If artificial neural networks can discover vision techniques that resemble biological ones, what other similarities could exist between the two vision systems?

Optical illusions cause the brain to see things that are not really there. Could artificial neural networks also fall victim to optical illusions? You might be surprised to learn that the answer is yes—and in much more bizarre ways than biological systems. Researchers have discovered that by understanding how neural networks are trained, one can generate images that can have the effect of fooling artificial neural networks. These are called *adversarial images*. We briefly discussed that artificial neural networks are trained by minimizing a loss function. That is, artificial neural networks are trained by modifying the internal connections so as to minimize the prediction errors for each iteration of training. Researchers realized that if the model parameters can be modified by the training process to minimize the prediction errors, then working with the opposite intention, where the model parameters are modified during training to increase the prediction errors, might yield interesting results.

For example, we could use the same principle to modify the input image in such a way as to maximize the error for each iteration of training. This can result in very interesting classification errors where a neural network might predict that an image that looks like random noise is a cat, an elephant, or any one of the categories it is trained to predict (fig. 2.7). Researchers have also discovered that some images can fool artificial networks of different architectures and training. A recent research paper found that some images that can fool artificial neural networks can also fool humans! To be fair, the researchers had to place a few constraints on the system to fool humans. These constraints help shed more light on the differences between human and artificial perception. For example, to fool the humans in the study, the images were flashed on a screen for a very brief period, and the human subjects had to quickly predict the category of the images they were seeing. The images could only be shown for a very brief period, or the humans would quickly realize the true nature of the images.

You see, the human visual system is much more sophisticated than our current artificial neural networks. For one, there are many more millions of neurons in the human visual cortex, with orders of magnitude more connections, than in artificial neural networks. The neural networks responsible for vision in humans are also not simple feed-forward networks like the artificial networks we have discussed. Feed-forward neural networks have connections between layers with information flowing in one direction from the first layer down to the last, hence "feed forward." The neural networks in our brains have connections that loop back between the layers and form recurrent loops between the layers as well. This has the effect of reenforcing certain features and makes our ability to perceive the world around us more robust. When trying to fool humans and artificial neural networks using the same images, the researchers minimized the effect of recurrent connections in our brains by letting the human subjects see the images for only a brief time. So what does this experiment teach us? What is the purpose of showing that images that can fool artificial neural networks can also fool humans?

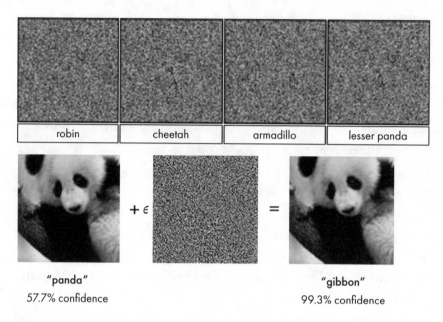

Figure 2.7 Two different kinds of artificial optical illusions created by different researchers that have been able to fool neural networks. *Top,* four examples of noisy-looking images that caused a well-trained neural network to predict "robin," "cheetah," "armadillo," and "lesser panda" for each sample. *Bottom,* an image of a panda: the trained neural network predicts that this is an image of a panda with 57.7 percent confidence; researchers then modify the image by adding a small amount of noise, resulting in an image that looks no different to us but causes the neural network to reclassify the image as a "gibbon" with 99.3 percent confidence. *Image of "panda"/"gibbon" from Ian J. Goodfellow, Jonathon Shlens, and Christian Szegedy, "Explaining and Harnessing Adversarial Examples" (poster presented at the Third International Conference on Learning Representations, San Diego, CA, 2015), https://doi.org/10.48550/arXiv.1412.6572. Image of noisy "robin," "cheetah," "armadillo," and "lesser panda" from Anh Nguyen, Jason Yosinski, and Jeff Clune, "Deep Neural Networks Are Easily Fooled: High Confidence Predictions for Unrecognizable Images,"* in 2015 IEEE Conference on Computer Vision and Pattern Recognition (CVPR), *427–36 (Washington, DC: IEEE Computer Society, 2015).*

Images that can fool neural networks of different architectures, including biological ones, suggest that when it comes to understanding vision and emulating it in artificial systems, perhaps we are really onto something. If the same image can fool biological neural networks and artificial neural networks that look vastly different, even when the artificial neural networks were trained using different loss functions, it might mean that the networks are all converging on a fundamental set of principles for interpreting the world. Sure, the rules they are discovering aren't perfect—hence the incorrect predictions—but it is exciting to see artificial systems emulating biological ones even when the results are incorrect.

○ ● ○

In this chapter, we set out to understand neural networks in the context of computer vision. We used a CNN as our conduit for comparing artificial and biological vision. We learned how features are extracted from an input image through each layer of the CNN down to the classifier. We now understand that artificial neural networks work by discovering the probability distribution of the features that make up their training data and use the learned distributions to make predictions about new samples.

It is encouraging and very exciting to see the similarities between biological and computer-vision systems—especially when the similarities are emergent, as if we suddenly, in our half-blind stumbling through an uncharted cave, happened to hit on a gold vein. These similarities, and the results we have achieved, encourage us to continue in the pursuit of the perfect neural network (is there one?). Today, artificial neural networks have been able to outperform humans in many specific categories, with "specific" being an important qualifier.

We can train a neural network to detect melanoma on a data set of 30,000 images from a hospital in Toronto, and that neural network will outperform a human pathologist by an accuracy of at least 10 percent against the same data set. But now grab a few samples from a hospital in Denmark that happens to prepare the biopsy slides in a slightly different manner (e.g., using different tissue-staining techniques), and the trained neural network that outperformed the human against the Toronto slides might be completely useless against the Denmark samples—yet the human pathologist will be just fine. Examples like these show us that, although we have come a long way in our pursuit of artificial vision, humans still see the world differently from neural networks. We are definitely better at generalizing concepts. When we teach a child a new word or a new class of objects—say we teach her what a soccer ball looks like—we do not need to present her with 10,000 images of soccer balls. Somehow her brain is capable of understanding the fundamental features of a soccer ball from a single sample. The child will (with some additional exposure to far fewer than 10,000 data points) be able to recognize any design, size, and color of soccer balls. Artificial neural networks are not there yet. It might be that simply modeling distributions based on a set of samples is destined to biases toward the training data, dooming the algorithm to excel only in specific settings. But research is ongoing, and if history is an indication, we will solve that problem too.

We can now answer the burning question of how similar computer vision is to human vision and, by extension, how similar artificial neural networks are to biological ones. Are artificial neural networks akin to some poor soul forsaken to a canned existence inside the Matrix? As of today, neural networks exhibit many elements also found in biological information processing, but at their core artificial neural networks are modeling distributions by minimizing a loss function. Minimizing a single loss function might be too simple an approach if our goal is to mimic the human brain. We saw the downfalls of minimizing the loss; the opposite is also possible. We can maximize the loss and get funny, bizarre results from our predictions. Our brains appear more robust at handling and processing input data.

In the next chapter, we dive deep into distributions and the decision-making process of neural networks. We learn more about loss functions and see, step by step, how neural networks can be trained using gradient descent through back propagation. This is, without a doubt, one of the most elegant algorithms in computer science.

3

ANSWERING AN
AGE-OLD QUESTION

We are slowly uncovering the secret powers of neural networks. The last two chapters described neural networks as funnels of information where the input data is compressed into a latent vector—a vector representation of the input. Then, the network performs either classification or some type of linear regression to get our output (fig. 3.1). In this chapter, we learn more about this last bit by discussing the mathematical assumptions and tools we use to classify or to analyze data samples to forecast some trend.

This is the chapter where we learn what makes neural networks tick. Our goal is still to find out what an AI system is truly *thinking*. This is important because it is intrinsically interesting and inspiring that a species not far removed from apes has created something that can be confused with intelligence. It is also important because anything new can be scary, and fears of AI have been explored and, to some extent, fed by Hollywood for decades (with movies like the *Terminator*, the *Matrix* series, and, more recently, *Ex Machina*). But concerns with AI and intelligent robots are not new and did not start in the latter part of the twentieth century; already in the 1940s, the science fiction author Isaac Asimov warned about the possibility of a machine takeover. As prophylaxis against such a scenario, he introduced the Three Laws of Robotics:

1. A robot may not injure a human being or, through inaction, allow a human being to come to harm.

2. A robot must obey the orders given it by human beings except where such orders would conflict with the First Law.

3. A robot must protect its own existence as long as such protection does not conflict with the First or Second Law.

More recently, AI entrepreneurs like Elon Musk have voiced similar fears of a Terminator-like machine rebellion if we don't constrain our current accelerated pace of AI research and adoption. AI researchers, on the other hand, seem less worried about machines deliberately deciding to take over the world and more concerned with the misuse of AI by humans, as discussed in *Scientific American*'s "What an Artificial Intelligence Researcher Fears about AI."

To really understand what is scary about AI and how we avoid dangerous consequences, we should first attempt to understand AI. In this chapter, we explain AI algorithms as probability estimators.[12] To do that, we learn about probability distributions and about linear and logistic regression algorithms. We also look more deeply at how neural networks are trained and how the model parameters are adjusted using calculus. You will not need to understand the math to get the message of this chapter, but it can be quite beneficial to at least try to understand the math. In the next chapter, we address the concerns around misuse of AI. Here, we discuss the fundamentals of how neural networks operate. After reading this chapter, you will have the tools to develop an informed opinion for the flavor of concern that makes sense to you.

---

12. For simplicity, I am using the terms *probability* and *likelihood* interchangeably, though there are technical differences between them.

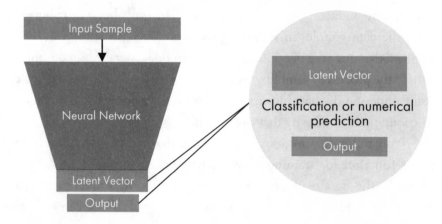

Figure 3.1 A neural network as a funnel of information. Information is input at the top and compressed through the network down to a latent vector representation. The output of the network is then typically calculated by performing linear regression (if predicting a continuous value) or logistic regression (if performing some classification).

Now that we have built a functional understanding of two of the most common classes of neural networks—the MLP (fully connected neural network) and the CNN (convolutional neural network)—we are going to dig a little deeper and peek into the soul of these algorithms. We have seen how a neural network consists of an arrangement of neurons with connections that are assigned special values, or weights. We understand that these weights are adjusted during the training phase of a neural network, and we know that these adjustments are where the magic happens. As we adjust the weights of the neural network, its predictions begin to get more and more accurate. And although we have skirted around the training process and briefly discussed a loss function as a measure of success, our discussion has been very superficial. Why is it that minimizing a loss function should cause a complex mathematical function (as discussed, we can consider a neural network a mathematical function) to map a set of input features to a class of desired outputs, especially for data samples it has never seen before? We have thrown around the term *probability distribution* and explained that neural networks are in fact learning the parameters of a probability distribution from the data samples in a training data set. We have said, or at least implied, that once the distribution is understood, generating an output is simply the process of sampling from the learned distribution. But do we really understand what this means? Do we understand what probability distributions are and why they are helpful?

Probability distributions are so crucial that the process of creating a loss function for training a neural network is the most important aspect in the design phase of a neural network. By now, researchers have identified loss functions that work for many of the use cases where we want to employ our neural networks, and most software engineers, when creating a neural network to solve a specific problem, just choose a loss function that has been proven to work for that class of problems. But identifying a new loss function or improving an existing one involves defining the problem we want the neural network to solve, in terms of probabilities. That is, we want the neural network to calculate the probability that a set of features for a given sample belong in each one of the categories defined by the network's outputs. This is the bit of magic that allows the neural network to work beyond the training data set. If we were to embark on defining a new loss function without considering how it could maximize the probability that the input features belong in the desired output category, we might find that our loss function produces weight values that lead nowhere or ones that only work for the training data set but do not generalize to real-world data. So, to better understand these important facets of artificial intelligence and its continuing development, let's dive into probability and its loyal companion, statistics.

At the heart of any algorithm that aims to predict an outcome for a data sample—for example, predicting the price of a house given some information about that house or predicting the species of a certain bird given the length of its beak—is probability and statistics. All the artificial intelligence algorithms we use today, including neural networks, use concepts from probability and statistics to make predictions on data. Why use statistics? Statistics is the best tool we have for making sense out of data. Everything we do in our lives generates data: withdrawing money from our bank, buying food from the grocery store, the type of food we buy, the amount of money we spend every month, and the movies we watch on streaming platforms. When information like this is kept together in a database and attached to an individual, it provides an excellent description of what that individual is like and their day-to-day habits. Today, data is one of the most highly valued commodities; companies like Facebook and Google earn billions of dollars off the information they maintain about you. And it is all thanks to probability and statistics. Statistics provides a toolbox for manipulating and extracting information from data. Based on information about events in the past, statistics helps build a timeline or relationship between those events. When we combine data with concepts from probability theory, we are able to predict future outcomes based on past events.

For example, most online retailers or even streaming platforms are well known to employ automated recommender systems aimed at suggesting new products you should buy or new movies or TV series you should watch. And you have probably noticed that, for the most part, they tend to be disturbingly good at suggesting items you would like. They can do this by building a *model* of you. By keeping a collection of data points over time—which items you buy, which movies you watch—they can predict what items you might be interested in. As we will see, designing a system to predict which shoe brand I am most likely to buy is not much different from designing a system to predict house prices or classify images of cats.

All artificial intelligence problems start with a data sample and a question about the sample. A data sample might be a single house in a data set of different real-estate listings. Let's say the house has four bedrooms. The question might be: "How much is the house worth, given the number of bedrooms?" If we have a database somewhere relating the number of bedrooms per house to property value, we could build an algorithm that is able to analyze the data in the database and extract the information we want to learn. What type of information might this be? Well, we would want our algorithm to learn a relationship between the number of bedrooms and property value. By learning this relationship, the algorithm could then apply this information to a brand-new house it has never seen and guess the price of the house based on the number of bedrooms of that new property. But we are getting slightly ahead of ourselves. To understand how artificial intelligence systems work, we need to understand probability distributions.

In this chapter, we focus on three well-known probability distributions: the *binomial distribution*, the *normal* or *Gaussian distribution*, and the *Bernoulli distribution*. If you haven't heard these terms before, don't worry; we cover them below. We do not get too deep into the mathematics, but it is important to understand a few basic concepts to make progress. We only go as deep as necessary to conceptually understand how we can make predictions about the future, based on information from the past. Here we go.

PROBABILITY DISTRIBUTIONS AND COIN TOSSES

What is a statistical model? A statistical model is a set of assumptions we make based on our current data, to help us make predictions about future data. The purpose of AI is to learn a set of good assumptions about existing data so that we can make predictions about future data. If this sounds like the definition of a statistical model, it is not a coincidence. So how do we make those assumptions? The answer involves probability distributions. Before we go any further, let's examine what we mean by probability distribution.

Suppose we conduct an imaginary survey to find out the average height of fifteen-year-old boys. Let's say that we go out and survey one thousand boys, and we find the distribution of boys per height as represented in table 3.1.

Table 3.1 Distribution of Boys according to Height in a Survey

Height (m)	Number of boys
1.64	30
1.66	80
1.68	200
1.7	400
1.72	220
1.74	50
1.76	20

Our distribution suggests that most of the boys in our survey are around 1.7 meters tall. If we build a histogram to visualize our data (fig. 3.2), we notice that the histogram resembles a rough bell shape, and it shows that most of the information is stored around the 1.7-meter mark, which happens to be the average, or *mean*, of our data. Our histogram shows sharp jumps between the different measurements in our survey. The sharp jumps are due to our survey being small (and imaginary!) so that although we can conceive that some boys might measure anywhere between 1.64 and 1.66 meters, we didn't find any in our survey. Using this information, we can also create a probability distribution graph that shows the probability that a fifteen-year-old boy measures any of the possible heights in our distribution (fig. 3.3).

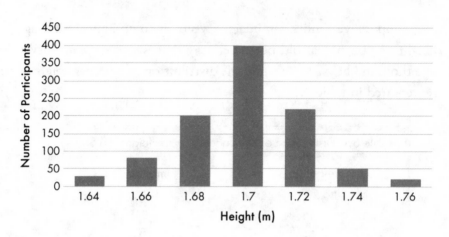

Figure 3.2 Histogram of one thousand participants in a survey, arranged according to height.

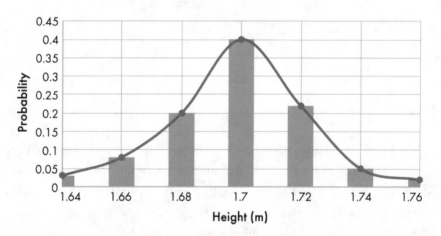

Figure 3.3 Probability distribution for each possible measurement (*orange bars*). The *blue line* shows the continuous probability for the height of the surveyed participants in the 1.64–1.76 m range.

To better represent the gaps in information that our survey might be suffering from, we could draw a continuous line that connects the edges of our histogram bins. When we do this, we can see a graph that tells us that most fifteen-year-old boys are on average 1.7 meters tall, and that lets us predict the probability of a fifteen-year-old boy being anywhere in the 1.64-meter to 1.76-meter range. A formal way of describing the information we have just seen is to say that figure 3.3 represents a probability distribution for a population of fifteen-year-old boys based on their height. In our case, the size of our population is one thousand boys. And the mean height of that distribution is around 1.7 meters. Why is it called a probability distribution? Because the graph shows the continuous probability that a boy is any height in the range of 1.64 to 1.76 meters. Another way to state this is that the graph shows how the probabilities of all possible height measurements for the surveyed participants are distributed in the range of 1.64 to 1.76 meters.

Why is this information important? Suppose I tell you that there is a boy who was not among the one thousand boys surveyed. All we know is that he is fifteen years old. If I ask you to guess his height based on his age, what would you guess? Given that you have seen our survey and data set, you would probably guess that he is 1.7 meters tall. If you did, then fantastic! You have just designed your first AI system. Figure 3.3 tells us that there are other possibilities for his height; 1.65 meters and 1.73 meters are both possibilities within our distribution. But given that we don't have any more information than his age, and we must pick a single value, picking the average is our best option, and it's a pretty good option. We might be off in a few cases, but if we have to predict the height of a group of boys, we can expect that by picking 1.7 meters we will be right more often than if we just randomly guess a height. Intuitively this feels right. It feels right because we are used to thinking in terms of average. It is natural for humans to make decisions based on averages of experiences. It even makes sense from an evolutionary point of view.

Imagine that we are back on the African savanna two million years ago. We have left the comfort and safety of the trees and descended its branches for the last time. Before us stand the open grasslands, full of new opportunities and new dangers. To succeed, it will help us to learn from the average of our collective experiences. If one of our friends runs down the field and gets eaten by a lion, and we immediately change our course of action based on this single case, our progress is going to be limited. To leave the trees behind, we must be able to accept new dangers, and we can't give up on a reward—crossing an important river or chasing a young antelope—simply because a single negative event has occurred. Similarly, if every one of our friends who tries to cross the river dies, and we don't adjust our behavior, we are not going to get very far. It is evident that there is a heavy cost to living on the edge of our distribution. One extreme is that we immediately change strategies as soon as we encounter a bad outcome (our friend is eaten by a lion). This is costly because, on the African savanna, bad experiences are going to happen often enough that if we change course every time, we are not going to make much progress. The other extreme is that if we keep encountering bad outcomes (our friends drowning crossing a river) and we do not recognize that we must change strategies, the cost is that we are all probably going to die, and we won't learn much in the process. A better approach would be to stay in the middle of the two extremes and adjust strategies only after a number of negative outcomes. Another way to say this is that a good strategy would be to continue doing the things that work most of the time.

What our ancestors were doing on the African savanna was building a statistical model using a learned probability distribution, just like we did to predict the boys' height using the information from our survey. We are going to come back to these concepts many times throughout the book because this is the essence of what every artificial intelligence algorithm does. Every artificial intelligence algorithm is trying to build a statistical model by learning the parameters of a probability distribution function. Different AI algorithms use different techniques to learn those parameters, and some are more successful than others, but they are all trying to build a statistical model. Let's look at another example of probability distributions, using a set of coin-flipping experiments. Once we understand probabilities, we will see how the last layer of a neural network leverages a loss function to learn the probability distribution for the data in a training data set.

For this experiment, we assume that we have a fair coin, where the probability of getting tails is 50 percent each time we flip the coin. We are going to define our experiment as consisting of ten flips. That is, a single experiment involves flipping a coin ten times. Next, we define a variable x as the number of tails we can get in each experiment; in other words, x refers to the number of tails we can get in ten trials for each experiment. Then, we are going to construct the probability distribution for each possible outcome of x.

We can ask what the probability is that we get zero tails in an experiment of ten trials (our ten flips). The probability that we get zero tails, which we can formally define as $P(x = 0)$, is the probability that every single flip in the ten trials lands heads. Out of all possible combinations of heads/tails in ten coin tosses, there is only one arrangement that produces zero tails. That is the case where all ten tosses land heads. The probability of this outcome is $\frac{1}{1,024}$. Remember that each trial only has two possibilities—heads or tails—and we are conducting ten trials in our experiment. This means that there are $2 \times 2 \times 2 \times 2 \times 2 \times 2 \times 2 \times 2 \times 2 \times 2 = 1,024$ possible ways to combine heads and tails in an experiment of ten coin tosses.

Let's illustrate what we mean with the following description of possible outcomes in ten coin tosses:

Possible outcome 1: HHHHHHHHHH → 0 tails in this experiment, all heads.

Possible outcome 2: THHHHHHHHH → 1 tail in this experiment on the first trial.

Possible outcome 3: HTHHHHHHHH → 1 tail in this experiment on the second trial.

Possible outcome 4: HHTHHHHHHH → 1 tail in this experiment on the third trial.

Possible outcomes 5–1,023: …

Possible outcome 1,024: TTTTTTTTTT → 10 tails in this experiment, no heads.

The above list of possible outcomes shows that—accounting for the fact that each trial can be a heads or a tails and knowing that our experiment consists of ten trials—we have 1,024 possible ways to mix heads and tails. Going back to our variable x, we can list the probabilities for each number of tails (our desired outcome) in a given experiment.

$P(x = 0) = \frac{1}{1,024}$ → Probability that we get 0 tails. There is only 1 outcome that can achieve this: HHHHHHHHHH.

$P(x = 1) = \frac{10}{1,024}$ → Probability that we get 1 tails. There are 10 outcomes that can achieve this (shift T to fill each position once).

$P(x = 2) = \frac{45}{1,024}$ → Probability that we get 2 tails.

$P(x = 3) = \frac{120}{1,024}$

$P(x = 4) = \frac{210}{1,024}$

$P(x = 5) = \frac{252}{1,024}$

$P(x = 6) = \frac{210}{1,024}$

$P(x = 7) = \frac{120}{1,024}$

$P(x = 8) = \frac{45}{1,024}$

$P(x = 9) = \frac{10}{1,024}$

$P(x = 10) = \frac{1}{1,024}$

If you don't quite understand how we can calculate the probabilities for each outcome in our experiments, don't worry; you just need to know that it *is* possible to calculate the probability for each outcome. And from those probabilities, we can build a probability distribution (as we will see next). If you are curious, however, as to how you can easily calculate those values without having to list all 1,024 possibilities and then counting each desired outcome by hand, the way to do it is to use *combinatorics*. Most calculators have a function that looks like *nCr*. To calculate the probability of getting six tails in ten trials, P(*x* = 6), we input 10C6, which equals 210. Then, we divide by the total number of possible outcomes (1,024). All right, now we can move on to building our probability distribution.

If we look at the orange bars of figure 3.4, we can see the *discrete probability distribution* for each outcome of *x*. It is called a discrete distribution because for each specific outcome, there is a single probability that describes the likelihood of that outcome. In statistics, the probability of different kinds of events occurring can usually be described by one of several well-known probability distribution functions. An experiment that consists of many trials and only has two possible outcomes per trial—success or failure (i.e., heads or tails)—can be described using the binomial distribution.

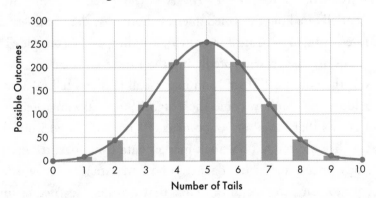

Figure 3.4 The probability distribution function for the coin-flipping experiment. The x-axis shows the number of tails we can get in an experiment comprising ten coin flips. The y-axis shows the probability of flipping any number of tails in ten trials.

THREE DISTRIBUTION TYPES

Here is why we have worked so hard to get to this point. It turns out that binomial distributions are very important in statistics because they are found to describe many processes in nature, so understanding binomial distributions can be a powerful tool in helping us understand nature. Remember, the purpose of artificial intelligence is to use the power of computer systems to extract information from data, to make predictions about future data. Probability distributions help us visualize the likelihood of each possible outcome in our experiments so that we do not have to conduct every experiment. To use our coin-flipping example: by knowing the probability distribution of possible outcomes for ten flips, we do not have to perform hundreds of experiments where we flip a coin ten times to empirically discover all the possible combinations of heads and tails. If we are asked to calculate the likelihood of observing six tails in an experiment of ten coin flips, we can just consult our probability distribution function. In general, if we know the likelihood of each possible outcome, then we understand our data enough to make predictions about future samples. This is the strength of probability distributions.

Why is this important in AI? Artificial intelligence algorithms are essentially making informed guesses about new data samples. Let's take forecasting algorithms for securities trading as an example. These algorithms are tasked with predicting the future price of a stock, but how can they do that? There is no magic ball that can tell us what the future looks like. So how could these algorithms make any predictions about events that have not yet occurred? What these algorithms have is data. They are trained on stock market data using past transactions and stock prices, and if the data is good and extensive, they can use this data to learn to recognize trends that can help to predict future market fluctuations and stock values. Imagine we know the probability distribution of possible prices for a given stock for any day of the year.

To predict the price of the stock for next Tuesday, all we have to do is sample from the distribution—that is, pick a price for the stock that, according to our probability distribution, is highly likely to be the stock price next Tuesday. To be clear: this is still a guess, but it's an informed guess. Artificial intelligence algorithms are constantly making guesses. When the algorithms are well trained, these are guesses based on highly probable outcomes.

In the medical field, artificial intelligence algorithms are used to analyze patient information to predict the likelihood that a patient is suffering from a given disease or will develop the disease at some point in the future. For instance, researchers are investigating the use of AI algorithms in the medical domain to predict the likelihood of a patient developing heart disease based on historical data. To do this, the hospital maintains data sets containing such patient information as age, gender, profession, height, smoking habits, history of heart disease in the family, and other information, along with an indicator of whether the patient suffers from heart disease. Similar to how we used the data from our survey to predict the height of fifteen-year-old boys, by analyzing the data from existing patients, the algorithm can learn relationships between each piece of information describing the patient and whether the patient has heart disease. Later, when a new patient walks in, a physician can ask the algorithm to predict how likely the patient is to suffer from heart disease given their age, gender, smoking habits, and other history. And it is all based on the algorithm's ability to learn the probability distribution for the likelihood of a patient developing heart disease for each possible combination of features for that patient (age, gender, smoking habits, etc.).

Whether we are trying to predict a person's height or determine the trajectory of a hurricane or classify images of birds into different species, there is a probability distribution that describes the underlaying data, which we can use to build a model. The binomial distribution is a discrete distribution because for each possible outcome there is a specific probability of that outcome occurring. The probability of getting only one tails in ten trials is a specific value. But not all distributions are discrete; some are continuous. In fact, we have already seen an example of a continuous distribution. In our example of predicting the heights of teenagers, we saw a continuous distribution. It is continuous because, although the mean of the distribution is 1.7 meters, no two boys from our survey are likely to have measured the exact same height. Consider that measuring someone's height is at best a good approximation. We have to account for the volume of hair on their head and how that might affect our measurement, the fact that not everyone stands perfectly straight, and so on. So a measurement of 1.7 meters is at best an approximation, with some boys being 1.7005 meters and others being 1.6999 meters. When the range of possible outcomes comprises continuous values, the distribution is a continuous distribution. The probability distribution from our heights example is called a Gaussian, or normal, distribution. The binomial distribution is the discrete version of a normal distribution. As we increase the number of trials in our experiment (if instead of ten coin flips, we perform one million coin flips), the binomial distribution will eventually approach the normal distribution (see the blue line in fig. 3.4).

The normal distribution is defined by two parameters, the mean (which we have already seen) and the *variance*. The variance is a measure of how much the data varies from the mean. A normal distribution with low variance means that the data is closely aligned to the mean; in other words, most samples in the data are close to the mean. On the other hand, a distribution with high variance means that the data is spread out away from the mean.

Knowing the parameters of a probability distribution is very powerful because it means that we can *sample* from the distribution. If we know the parameters of our distribution, we can pick values that conform to that distribution to generate new data. It also means that we can evaluate new data by seeing where it falls in our distribution.

Imagine we are taking part in a trivia night, and we are asked the following question: "John M., born on May 2, 1976, won MVP in which major league sport?" One of our friends at the table thinks he knows the answers. "It must be hockey!" he says. But before you let him answer, someone else at the table happens to be a statistician who did a study on NHL players, and she remembers two key pieces of information: the mean and variance for the probability that NHL players are born on a given day of the year. And she knows that the mean of that distribution is centered on March 3 with a variance of twenty days.[13] With this information she can tell that most NHL players are born between December 14 and March 23. She still can't answer the trivia question, but she knows the major league sport in question is likely not hockey.

This is precisely what artificial intelligence systems do all the time. They make educated guesses about the probability distribution behind a given process and then proceed to discover the parameters of that distribution by analyzing existing data. Once the parameters of the distribution are discovered, new data can be evaluated against the known distribution. In statistics, we have many probability distribution functions that can be used to describe different processes. We have the binomial distribution, the Poisson distribution, the normal distribution, the Bernoulli distribution, and so on. Each distribution has a different set of parameters that describes that distribution. And different problem types are better suited to different distributions. The Poisson distribution is good for describing counts or incidents in a defined period. For example, if we want to predict the number of visitors to our website over the next month considering the number of visitors we received each of the previous twelve months, we can assume the number of visitors each month is distributed according to the

---

13. This is a fictional example to illustrate how understanding the probability distribution of some process can help us make predictions. We don't know if the process describing birthdays of hockey players is Gaussian, and we certainly don't know the mean and variance of that distribution.

Poisson distribution. We can find the parameters of the distribution by analyzing how many visitors came to our website each day of the past year, and we can sample from that distribution to make predictions about how many visitors we will get over the next month.

We revisit the Bernoulli distribution later when we discuss classification algorithms. It is used to describe the probability distribution of data samples falling into one of two categories: dog vs. cat, orange vs. apple, heart disease vs. healthy. The normal distribution, which we have seen, is one of the most assumed distributions in artificial intelligence. Many processes—like population height, customer satisfaction, and birth weight—can be modeled using the normal distribution. The exact reason why this is the case is beyond the scope of this book but is explained by a mathematical theorem called the *central limit theorem*. This theorem states that whenever a population sample is large enough (a population sample is a subset of the population that we use in our experiments; e.g., the one thousand boys we surveyed are a sample of the overall population of fifteen-year-old boys in the world), the means of different population samples follow a normal distribution. This is powerful because it implies that whenever we have a process that depends on the aggregate of many subprocesses, the aggregate process can be modeled using the normal distribution. And many things in nature are the result of the sum of many other processes; for example, the birth weights of newborn babies are the result of many subprocesses (the size of the parents, different health markers from the mother, which is itself affected by different pressures for the society in which the mother lives, etc.).

At this point, we should understand distributions and the power of using probability distributions to make predictions. We have developed an understanding of how we can use statistics and probability to create a simple model from existing data. But what if our data sets are slightly more complicated? What if, instead of surveying only fifteen-year-old boys, we surveyed teenage boys between the ages of thirteen and nineteen and wanted to build a model that could predict their height in that age range? We could still assume a normal distribution and find the mean height of boys in the thirteen-to-nineteen age range, but you can intuit that the mean for that distribution will not be very useful to predict the height of a seventeen-year-old boy or a fourteen-year-old boy because the mean would be spread out over the entire thirteen-to-nineteen age range. What we need in this case is a method that can find the probability distribution of height for each age group in the thirteen-to-nineteen age range.

LINEAR REGRESSION: THE CONCEPT

Calculating the mean height of the sample population of fifteen-year-old boys was a good step for predicting the height of fifteen-year-old boys outside of our survey data set because the mean is one of the parameters of the normal distribution, and we can use it to measure where those boys fit in that distribution. But what if our data set consists of a range of ages, as we postulated above? In this example, we might still be able to calculate the mean height for each age group in the sample population and store those values to make predictions in the future, but that would only work if our data set comprises many samples for each age group from thirteen to nineteen. If our data set does not have samples for ages fourteen and fifteen, how could we calculate the mean height for these groups? In this case, our simplistic model would fail. When we apply artificial intelligence to solve problems in the real world, our data sets are far from perfect, and there are population groups for which we have no information or very little information. To solve this problem and approximate an answer for data that might be sparsely populated, we can use linear regression.

Linear regression is a widely used algorithm for predicting market valuations in financial institutions. It can be used for modeling customer satisfaction in the hospitality industry or predicting the rate of infection for a contagious disease over a population. In general, it is a method used for finding a relationship between a dependent and an independent variable. As we will see, linear regression is an algorithm in its own right, and it can be used without a neural network; however, as we saw in chapter 2, what many neural networks do to forecast information in their last layer is a form of linear regression. Using our house prices example, the independent variable is the number of bedrooms of the house, and the dependent variable is the house price. The price of the house is a dependent variable because our assumption is that the price of the house depends on the number of bedrooms of a house. Linear regression is performed by finding the line that best fits the data in our data sets. The best-fit line is the line that minimizes the distance between the points on either side of the line and the line itself (fig. 3.5). It is a powerful device for finding relationships and making predictions about data points where there is missing information, or gaps in our data set. Let's take a look at a few illustrations of scatter plots and best-fit lines.

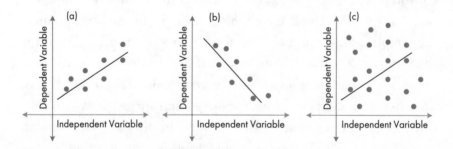

Figure 3.5 Three scatter plots with best-fit lines (*black lines*) running through the data points (*blue dots*). Subplot (*a*) shows a positive linear relationship between the dependent and independent variable. Subplot (*b*) shows a negative linear relationship between the dependent and independent variable. Subplot (*c*) shows data that does not exhibit a linear relationship between dependent and independent variables.

If we look at figure 3.5, we see three different examples of scatter plots with a best-fit line running through the data points. Scatter plots are used to visualize the relationship between samples in a data set. If we have a data set consisting of many samples, we can plot the samples on a graph where the x-axis represents the independent variables (in our house example, these would be the number of rooms in the house), and the y-axis represents the dependent variable. Figure 3.5*(a)* shows a *positive* relationship between the independent and the dependent variable. As the independent variable grows—that is, as the values increase in the x direction—the values also increase in the y direction. Subplot *(b)* shows a *negative* relationship between the independent and dependent variables. As values increase in the x direction, the values in the y direction decrease. From subplots *(a)* and *(b)*, we can see that the relationship between the independent and dependent variable is approximately linear. As the values increase or decrease in the x direction, the values in the y direction increase or decrease by the same factor. Understanding the relationship between the independent and dependent variables in our data set is crucial to understanding whether we can hope to build a model to interpret our data.

Suppose we have three separate data sets that produce scatter plots *(a)*, *(b)*, and *(c)*. In subplots *(a)* and *(b)*, we can see that the independent and dependent variables have a linear relationship. Remember that the dependent variable is the value we want to learn how to model (i.e., how to predict) given the independent variable. In these cases, we might consider using a linear regression algorithm to find the line that best fits the data. Once we find the best-fit line, that line becomes our model. Using this line, we can analyze data points outside of our data set and "predict" the value on the y-axis (i.e., the price of the house) (fig. 3.6).

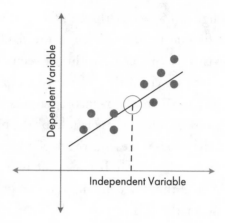

Figure 3.6 How a linear regression model (i.e., a best-fit line) can predict information for missing data in the data set. The *segmented line* indicates a value for the independent variable that does not correspond to any known value in our data set (visualized as a gap in the *blue dots*). The best-fit line can be used to approximate the missing information.

In figure 3.6, we see an example of a gap in the "knowledge" of our data set. If we are asked to predict the value of a dependent variable (maybe the price of a house or a person's height) given the value of the independent variable (the *x* value), we can't use the first approach we explored where we calculated the mean of the values in the data set and used the mean as our prediction. As we can see, we don't have any samples in our data set (blue dots) where there is a *y* value for the *x* value we are asked to predict, so we can't calculate the average of *y* values given our *x* value. By using linear regression to find the best-fit line, we can use the equation of that line to find the *y* value for our best-fit line at point *x*, and this can become our predicted value.

Now consider subplot *(c)* of figure 3.5. The scatter plot in *(c)* does not show an obvious relationship between the dependent and independent variables. If there is a relationship, it is definitely not linear. We can still use linear regression to find a best-fit line that runs through the data. The problem is that, since there is no linear relationship in the data to begin with, we would find the best-fit line to be a very poor model for making predictions about our data. Remember that the purpose of AI algorithms is to build a model from existing data to make predictions about data outside of our data set. If a new house comes on the market, we want to be able to predict the price of the house based on the number of bedrooms even though we have never seen this house. But if the data set we used to build our model looks like subplot *(c)*, the predictions we would make using that model would not be accurate enough to be useful. Even a cursory look at figure 3.6 shows that the predicted *y* value from the selected *x* value (the purple line's x-intercept) would not be a good prediction. We can see that the line does not accurately predict the existing data points in the scatter plot.

This is why good data scientists spend a lot of time understanding the data they are working with before running the data through a particular algorithm to create a model. Creating a model is easy. The trouble is in understanding whether the model we are creating is a good model to interpret our data set. In the subplot *(c)* example, a linear regression model would not be a good model to interpret that data. The term *artificial intelligence* can be misleading because it suggests there is some secret intelligence sauce inside a computer, and if we just shovel enough data into a computer and click the Artificial Intelligence button, the computer will sort it all out and give us the right answer. Hopefully, we are beginning to see that, instead of hands-off intelligence, AI is a collection of tools based on statistical assumptions and probabilities; to ensure that our projections work and are accurate, we must understand the data we are working with. Only then will we know whether the assumptions we are making make sense. Throughout the rest of the book, we will continue to see the pitfalls of failing to understand the data before selecting our algorithms or, worse, before deploying our models.

LINEAR REGRESSION: THE ALGORITHM

At this point, we should have a good intuition for what a linear regression algorithm does and how powerful it can be. We are now going to look at how linear regression can find the best-fit line given a data set and how probability distributions are related to linear regression.

First, recall that the linear regression model is a best-fit line running through the data in our data set. How can we create an algorithm, or a systematic set of steps, that results in a best-fit line? The first step is to choose a random line through our data set and then progressively adjust it. You are probably thinking, "A random line, really? How do we even choose a random line?" We can see from figure 3.7 that there are many possible ways to draw a line. In fact, there are an infinite number of possibilities for where we can draw a line in that graph. The answer is that we just pick one. We randomly select any line as our initial model.

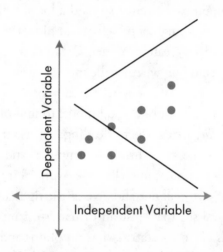

Figure 3.7 The two *black lines* are simply two possibilities among infinite ways to randomly place a line over a data set.

So how do we go from a random line to a best-fit line? It starts with the equation of the line. This is where all the work we have done so far starts paying off. In high school, we learned that the equation of a line is $y = mx + b$. In that equation, m is the slope of the line, and b is the y-intercept. Using those two parameters (m and b), we can generate any line we want, in the exact orientation we want it. By adjusting the m parameter, we can rotate the line in any direction, and by adjusting the *b* parameter, we can move the line up and down our coordinate system.

To create a linear regression model, we simply need to create an algorithm that can learn the *m* and *b* parameters of the line that best fits the data in our data set; and to learn those parameters, we use our training data. Before proceeding any further, there is a slight terminology tweak that we must make. In AI literature, the parameters of a model are called *weights* and *biases* and are typically denoted using θ. The input samples are denoted by *x*, and the labels are denoted by *y*. So, to put it in proper AI terminology, the equation of our line becomes $\hat{y} = \theta_1 x + \theta_0$, where $\hat{y}$ is the predicted label, as in the label that our new model has predicted for the input sample with features *x*. Note that $\hat{y}$ is different from *y*, the actual label (the "ground truth" mentioned in the last chapter).

We begin training our algorithm by choosing random values for θ_1 and θ_0. This gives us our initial random line. This line will not be very useful, but that's OK because we have to start somewhere. Next, we choose a sample from our training data set, replace *x* in the equation with that sample's feature value, and calculate a value for $\hat{y}$. The value for $\hat{y}$ we have just calculated is our predicted value for this sample. So, continuing with our house price data set, if the house in our data set has three bedrooms, then $x = 3$. θ_1 and θ_0 are initialized to random values, so let's assume $\theta_1 = 10$ and $\theta_0 = 5$; then $\hat{y} = 10(3) + 5 = 35$ (for simplicity let's assume that these are in $1,000 units, so 35 really means $35,000). Our linear model has predicted that our three-bedroom house is worth $35,000. I don't know where in the world you live, but I live in Toronto, and $35,000 for a three-bedroom house in Toronto is laughably inaccurate, which makes sense because we chose random parameters for our line. Now comes the fun part. Because we are training our model, and because we have a training data set, we have a label *y* for this sample that serves as our ground truth for the sample. So let's assume this house is valued at $1,000,000; then $y = \$1,000,000$. The job of our training algorithm is to compare our predicted value of \$35,000 with the ground truth of \$1,000,000, calculate how far our prediction was from the ground truth, and, based on this calculation, adjust the θ_1 and θ_0 parameters so that next time through our data set, the prediction will be closer to the ground truth. This is the essence of how all AI algorithms, including neural networks, are trained. We have already seen this in the previous chapters. We start with random parameters for our model. We generate predictions. We compare how far our predictions are from the ground truth, and we adjust our parameters so that next time through the training process, the predictions are closer to the ground truth.

When we are training an AI algorithm to build a model, the training process goes through the entire data set several times. Recall that in AI literature, when the training process goes through the entire data set once, this is called an epoch. Training AI algorithms to achieve robust results may take hundreds of epochs. This means that the training process presents each sample in the training data set to the model hundreds of times before the model is fully trained. Each time a sample is presented to the model, a prediction is made, and after comparing against the ground truth, the loss for the sample is calculated. In previous chapters, we saw that the loss is the difference between what the actual value is (the ground truth, or sample label) and the predicted value. The goal is to reach the point where the model's predictions closely match the ground truth. How do we know how many epochs it will take to train our model? We will get to that soon; for now, our focus is understanding how the model's parameters are adjusted through training.

Let's assume that our data set consists of one thousand samples (which is a very small data set to train a model, but we need to keep it simple while we are still learning). A single training epoch involves presenting each sample to the model one at a time. In a data set of one thousand samples, a single training epoch consists of analyzing one thousand samples, comparing predictions to ground truths, and calculating the loss for each sample. We then want to use this loss as a guide for how to adjust the model parameters to minimize this loss in the next epoch.[14]

To calculate the loss, we use what is called a loss function (as mentioned in chapter 2). The loss function is simply an equation that can help us calculate the difference between the predicted value and the ground truth in a way that can point us in the direction to adjust the model parameters. The loss function we select depends on the problem we are trying to solve and the model we are trying to train. Finding good loss functions for any given type of problem is an active area of research, and far beyond the scope of this book. For our purposes, all we need to know is that once an AI engineer has defined the problem and chosen a model to train (e.g., a linear regression model), there are well-known loss functions for different classes of problems that can be used to train the model. For linear regression algorithms, a common loss function looks like this: $L = \frac{1}{2}(h_\theta(x) - y)^2$. It looks complicated, but it really isn't. We know that y is the sample label, and x refers to the sample features. We also know that θ refers to the model weights. The only term that we have not seen is $h_\theta(x)$, and it refers to the model itself. Remember that AI models are considered mathematical functions. The term $h_\theta(x)$ describes a function that processes x features, and it is parameterized by weights θ. We use h as the name of the function instead of the traditional f because we call our model a *hypothesis*. This equation is simply saying: Calculate the difference between the predicted value $h_\theta(x)$ and the ground truth y, and square the result. We don't have to worry about the $\frac{1}{2}$ constant.

---

14. Again, I'm simplifying for ease of explanation. In practice, many neural networks use what are called *mini batches*, where samples are not presented one at a time but in groups of samples from the larger set. Instead of calculating the loss of each sample and updating the weights based on this calculation, we divide an epoch into small batches of samples. For example, for a data set of one thousand samples, we might create mini batches of one hundred samples. In this case, the epoch lasts for ten mini batches, where the loss is calculated for each sample in the mini batch, and the weights of the model are updated based on the loss over the entire mini batch. Updating the weights based on the loss over the mini batch instead of per sample makes the training process more likely to converge. Updating the weights based on the loss for each sample means that noise in the data can greatly affect the training process.

It doesn't affect the loss, or the point we are trying to make; it is simply there to make further calculations simpler.

You might be wondering why we even need a complicated loss function. And why do we need to square the difference between the predicted value and the ground truth? Why can't we just calculate the difference between the predicted value and the ground truth and be done with it? (Technically, this would still be a function, although admittedly a simple one.) The simplest answer is to say that squaring the difference helps us deal with negative values. For example, if the predicted value is smaller than the ground truth, the difference would result in a negative value. Squaring the value helps us get rid of the negative. The more accurate answer is that loss functions are carefully constructed because they must be differentiable, and they must be formulated to maximize the likelihood of predicting the correct outputs based on input features belonging to the same distribution as the training data. In other words, it helps to frame the problem we are trying to solve in terms of probabilities. Specifically, we need to show (mathematically) that by minimizing the loss function, we maximize the likelihood of predicting a correct output for a given input sample. When we frame the linear regression problem in terms of probabilities and calculate a method to maximize the likelihood of predicting the correct outputs based on input features, we end up with $L = \frac{1}{2}(h_\theta(x) - y)^2$. But more on this later.

Once we have selected a loss function, to systematically adjust the model parameters in a way that minimizes the loss over the training epochs, we need to use derivatives. That's right. We can finally answer the question that high schoolers have been asking for ages: When will I ever need to use calculus? To adjust the model's weights, most AI training algorithms, including neural networks, use derivatives. We explain this process next by trying to understand the following equation:

$$\theta = \theta - \alpha \frac{\partial L}{\partial \theta}$$

If we can control our survival instincts pushing us to flee the sight of an equation, we will see that this is in fact a simple concept. The first thing we need to do is identify the elements we are familiar with. We know that θ refers to the model weights. In a neural network, these are the weights of the connections between each neuron. In a linear regression algorithm, θ is the slope of the line we are trying to find.[15] L is the loss function. These are elements that we have already seen. Before we get into what α or ∂ means, we can deduce that the equation is trying to show us how to calculate a new weight. In other words, in the case of linear regression, it is telling us how to update the slope of the line we are trying to find. The equation shows us that to calculate the new value for the θ parameter (i.e., to calculate a new slope for our line), all we have to do is take the current value for θ (if this is the first iteration; remember that this is a randomly chosen value) and subtract the derivative of the loss function with respect to the current θ value. That is, the right side of the equation is updating the left side of the equation. The θ value on the right of the equal sign represents the current slope of the line, and the θ value on the left side represents the updated value for the next training iteration. ∂ is simply the sign for partial derivative. Let's try to unpack why derivatives help us update our model parameters.

First, let's remember what an equation is really telling us. Consider the following equation: $y = x$. An equation shows us the relationship between independent and dependent variables. If we were to plot a graph of the value of y given any value of x for the equation $y = x$, we would get the graph shown in figure 3.8. A different equation, such as $y = x^2$, will show a different line, in this case a parabola (fig. 3.9).

15. Note that we must perform the same calculations for each θ in a model (i.e., θ_1 and θ_0). To make our explanations easier to follow, we will generally ignore the y-intercept (i.e., θ_0) and focus on the slope (i.e., θ_1). Just remember that we must perform the same calculations for every weight in a model (in a neural network, this can mean millions of θ values), and this includes the biases.

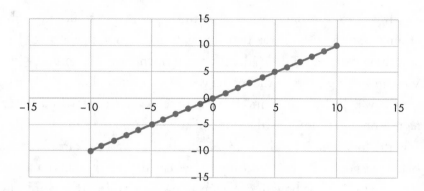

Figure 3.8 A plot of the line $y = x$.

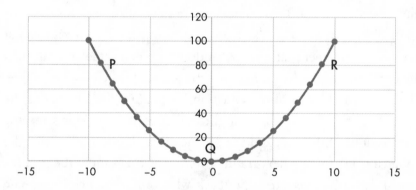

Figure 3.9 A plot of the line $y = x^2$.

Regardless of type, the line produced by an equation describes the range of possibilities for the dependent variable (y) given any possible values for the independent variable (x). When we are dealing with a simple equation of a single independent variable x, our equation describes a one-dimensional world. This is why the graph of the equation is a line. And this is the case in simple linear regression where the data samples are described by a single feature. It does not have to be a straight line, but it will be a line. If the equation is slightly more complicated and includes two independent variables, such as $y = x_1 + x_2$, then the equation describes a surface in 3D space—remember our vectors in N-dimensional space from chapter 1, where we described a vector consisting of a line originating at the origin and traveling through an N-dimensional space where x_1, x_2, x_3 describes how much the vector should move in each dimension. This surface would not look like a line; it would look like a topological surface that describes the possible values of y considering all the possible values of the x_1 and x_2 dimensions.

When dealing with AI, where our input vectors deal with multiple dimensions (i.e., x_1, x_2, . . ., x_n), our equations describe a multidimensional surface of possible y values. In these cases, linear regression is still finding a best-fit line through the data, but it's a line in a multidimensional surface. It's important to understand that, while we have been using a simple example where the data samples contain only a single feature (the houses in the data set are only described by one dimension, the number of bedrooms), data samples are invariably described by multiple features (the number of bedrooms, location, proximity to hospitals, etc.). The multiple features add extra dimensions to the equations that AI algorithms need to solve. We are using a single feature to simplify the explanations, but the algorithms are the same whether we are dealing with one feature or multiple features. The math is just easier with a single feature.

Now let's go back to our loss function, $L = \frac{1}{2}(h_\theta(x) - y)^2$. This equation is describing all the possible loss values our model can produce for all possible input samples and all possible parameters θ. If we close our eyes (you might have to finish reading this sentence before closing your eyes), we can try to imagine a surface floating in space, and this surface has some topological features like mountains, planes, and valleys. The mountains are areas where the loss is really high, and the valleys are areas where the loss is really low. We can think of the valleys as describing the times (combinations of θs and inputs) when the model makes the correct prediction (that's why the loss is low) and the mountains as the times when the model makes incorrect predictions, causing the loss value to be high. Our goal in our training algorithm is to traverse this terrain in search of the valleys. Let's look at the equation again: $L = \frac{1}{2}(h_\theta(x) - y)^2$. Considering that the input features (x) and the ground truth (y) cannot be adjusted, the only variable we can alter to ensure L remains low is θ. That is, our goal is to adjust θ so that the loss for each sample belongs in the valleys of the loss surface, not in the mountains. The question then is, How do we adjust θ in a direction that moves us toward the valleys of the loss function?

Instead of a multidimensional surface that we can't even picture, let's assume our loss function describes a line. Figure 3.10 shows what this surface might look like. We can now assume that during our training, when we calculate the loss L for a given sample, depending on what the features x and the θ parameters are for that sample, the loss will be in some place along that curve. Let's consider what happens during the first iteration.

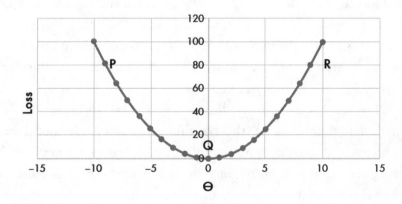

Figure 3.10 What a loss surface for $L = \frac{1}{2}(h_\theta(x) - y)^2$ might look like if the input data sample contains a single feature (i.e., x_1).

We randomly initialized our parameter θ, so let's assume that after calculating the loss for the first training epoch, our calculated L is at point P in figure 3.10. Point P is high on the curve because our loss is large at this point. Our goal is to find a systematic way such that every training step we take moves from point P closer to point Q. Why do we want to move to point Q? Because point Q is the lowest point in the loss surface. Instead of simply accepting what I have just said, try to visualize what this means. The loss surface describes all possible loss values we could encounter during training. When the loss is high, our predictions are very far from the ground truth. When the loss is small, our predictions are very close to the ground truth. And the goal of training is to drive our predictions closer to the ground truth. Another way to think about this is that we want to traverse the loss surface from the high place where we find ourselves to the lowest point in that surface. To do this, we (again) use calculus.

Remember that calculus helps us measure rates of change. Let's go back to our equation for adjusting the model parameters:

$$\theta = \theta - \alpha \frac{\partial L}{\partial \theta}$$

To solve this equation, we need to calculate the partial derivative of the loss function with respect to parameter θ. This calculation tells us how changes to θ affect the loss L. Another way to say this is to say that for some loss P, we want to figure out how adjusting θ affects the loss: Does increasing θ also increase the loss to some point higher than P, or does it decrease it? Going back to figure 3.10, $\frac{\partial L}{\partial \theta}$ tells us how changes in θ move P farther away or closer to Q. One way to think of this is that $\frac{\partial L}{\partial \theta}$ is the slope (in a multidimensional surface, we call this a *gradient*) of the L surface at point P. To move from P to Q, we want to calculate the slope of the curve. The slope tells us the direction in which we want to move. Here is why: At point P in our graph, the slope of the curve will be negative, so $\frac{\partial L}{\partial \theta}$ will be a negative value. Note that to calculate a new value for θ, we take the current θ and subtract the slope of the surface. But if that slope is negative, then $\theta - (-\text{slope})$ becomes $\theta + \text{slope}$. So the new θ value will be slightly larger. If we look at figure 3.10 then, increasing the value of θ is exactly what we want to do to get to point Q. As θ increases, the value of L decreases, and the update nudges us in the downward direction toward point Q.

But what if instead of finding ourselves at point P originally, we found ourselves at point R? Then, the slope $\frac{\partial L}{\partial \theta}$ of our curve at point R would be a positive value. To calculate a new value for θ, we calculate θ − slope. This has the effect of decreasing the value of our new θ. This makes sense because, going back to our curve, to get to point Q, we want to decrease the value θ, which decreases the value of L. In chapter 1, when we described updating the model weights as adjusting some tuning dial, this is exactly how we turn those dials.

This process of calculating the gradient (the slope in a multidimensional surface) of the loss surface and moving in the downward direction of the gradient toward a minimum loss value is called *gradient descent*. The state-of-the-art method today for training artificial intelligence models and updating the model parameters to minimize a loss is gradient descent. In a neural network, the model is composed of multiple layers, and each layer has many neurons with connection weights that must be updated. The process of updating those weights is the same as what we have just described. We must calculate the partial derivative of the loss function with respect to each weight to update each weight value. Since neural networks are composed of multiple layers and we must perform this calculation for each layer from the output layer all the way back to the first layer, we call this process *gradient descent through back propagation*.

I find gradient descent through back propagation to be one of the most elegant solutions in computer science. To understand its significance, it helps to remember that none of this was obvious. For decades, neural networks seemed impractical because there wasn't a good method for updating the weights on large networks. Someone had to come up with the idea of using calculus in this way and have the conviction to try it.[16] And to think that this actually works should give us hope and excitement for what we can achieve!

Going back to our linear regression model, by adjusting the θ parameter systematically so as to minimize the model's loss, what we are really doing is adjusting the slope of the best-fit line. Recall that our linear regression model is simply a line with equation $y = \theta_1 x + \theta_0$. When the slope of the best-fit line was randomly chosen, the predictions were bad because the line did not accurately represent (or fit) the trend in our data. But as we adjust the slope of the line, the predictions become better as the line eventually locks in on the trend.

PROBABILISTIC INTERPRETATION OF LINEAR REGRESSION

We started this chapter discussing probability distributions, and we showed how understanding them can help us identify trends in data, which helps make predictions about new data. We saw that if we can assume that the data we are analyzing is distributed according to some known probability distribution, then we can try to discover the parameters of that distribution. Now we are going to see how linear regression is related to Gaussian distributions by describing a probabilistic interpretation of linear regression. We have already seen the geometric interpretation of linear regression, and we know how to train a linear regression model on a data set to make predictions. Why do we need to understand the probabilistic interpretation?

---

16. It is difficult to pinpoint who "invented" gradient descent through back propagation because, as is often the case, many people over decades came up with contributing ideas, each building on previous work. Yann LeCun, currently chief AI scientist at Meta, is often credited with creating the first practical implementation of back propagation in 1989, while working at Bell Labs. Arthur E. Bryson and Yu-Chi Ho are credited with inventing back propagation to train deep-learning models in 1969. But it all depends on how we define "invented" and where in history we draw the line.

The purpose of this book is to help us understand the elegance of artificial intelligence systems—and their limitations. The probabilistic view helps guide our intuition for how accurate our predictions are and how much we should trust our system. We need to remember that the purpose of artificial intelligence is to help us build models that can make predictions we can use to then make decisions about the future. If we consider these AI systems magical "black boxes" with inherent *intelligence*, then we are setting ourselves up for disappointment when their predictions fall short of reality. This is what happens when modeling systems make predictions about the weather and, instead of sunshine, we see rain; or more recently, when modeling systems predict a fall in COVID-19 infection rates, but instead the infection rates increase. In these cases, most people blame the systems as useless and reject them altogether. Once we understand how these systems work, we see that whether the predictions prove accurate or not says little about the system's intelligence. Predictions are based on the system's ability to model a distribution. If the system was trained using a data set that represents a good distribution of outcomes, then the system should be able to make accurate predictions. If the system was trained using a data set that isn't very representative of reality, however, then the predictions will not be very accurate.

For example, suppose we trained our linear regression model to predict the height of fifteen-year-old boys again. But for the training data set, we chose one thousand boys all belonging to different basketball clubs. As we might imagine, in this case the algorithm will be biased toward tall boys, since presumably basketball players are very tall (or at least taller than average). If we then ask it to predict the height of a new fifteen-year-old boy who walks in off the street, the prediction might prove inaccurate. If, instead, our data set is chosen from a random population of fifteen-year-old boys from the city of Toronto, our algorithm might prove more accurate at predicting heights of fifteen-year-old Torontonians. In both cases, it is the same system; in one case it succeeds, and in another it fails; and it all depends on the training data. (So we may well ask, Where exactly is the intelligence?) Note that while our algorithm might prove accurate at predicting the height of random fifteen-year-old boys in Toronto, that same trained algorithm would prove highly inaccurate in other cities around the world. This is the problem of bias, which we discuss more in the next chapter.

So what does linear regression have to do with probabilities? We discussed the method for training a linear regression model, and it involved calculus. Where do probabilities factor in? Probabilities factor into the loss function $L = \frac{1}{2}(h_\theta(x) - y)^2$. We are not going to derive the loss function because that process is quite complex and beyond the scope of the book, but we will intuit how probabilities lie at the heart of our training algorithm. First, let's consider a hypothetical data set where the data lies exactly on our predictor line. And I don't mean that just our training data set will lie on the line (see fig. 3.8). I mean that we are guaranteed that all the data, including the real-world data, will lie exactly on the line. That is, for every data sample, $y = \theta x$. In this case, we don't really need gradient descent, do we? All we need to do is pick two data points and calculate the slope of the line that runs through them. Since the data points line up perfectly on the best-fit line, the slope of the line between any two points will perfectly describe the line for all points.

The reason we need gradient descent is because the data, although it can exhibit a linear relationship, contains noise. This noise means that the data points won't exactly lie perfectly on our best-fit line. If we were to pick two random data points, calculate the slope of the line between them, and use this as our predictor, the predictor might not be very accurate since the line running through any two points will not necessarily result in a best-fit line (fig. 3.11). It is because of this noise that we use gradient descent to find the best-fit line.

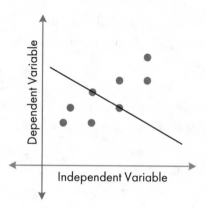

Figure 3.11 A line running through two random points in the data set. The *black line* is clearly not a best-fit line, yet it runs through two points.

Gradient descent explains how to update the model weights by minimizing the loss, to improve our results. But it says very little about why this works or why we have used a specific loss function. Suppose we grabbed a random mathematical function and used it as our loss function. Would that work? Most likely it would not. To develop a loss function effectively, we typically start by expressing the problem we are trying to solve in terms of probabilities. In our case, we want to calculate the probability of a data sample having label y given a set of features x, using a model with parameters θ—mathematically, we express this as $P(y|x; \theta)$. We have already said that our data samples contain some noise. We will call this noise ϵ and can express it more formally as $y = \theta x + \epsilon$. Note that $y = \theta x$ describes a line where the data samples lie exactly on the line, which can be thought of as $y = \theta x + \epsilon$ when ϵ is 0. When ϵ is not 0, there will be some "jitter" in the data, sprinkling the data points about the line.

Here we assume that this noise is distributed following a normal distribution. Recall that, according to the central limit theorem, this is often an OK assumption to make for observations that depend on many processes. So although we don't know the noise for each data sample (i.e., we don't know exactly how far each data sample will be from the line), we assume that the noise follows a normal distribution. Armed with this information, we can then discover the parameters of that normal distribution: the distribution's mean and its variance. The mean and the variance of the distribution then allow us to sample from that distribution, which is essentially what the best-fit line is doing. We can think of the best-fit line as sampling from the distribution of noise in our data, resulting in a line that runs as close as possible to all the data points. If we restate the linear regression problem as calculating the probability that for each data sample, the features x belong to label y, or $P(y|x; \theta)$, we can start by writing the equations in terms of the normal distribution equation. Then, we can proceed to calculate the maximum likelihood that features x belong to each label y. This process can fill several pages with algebra; I will spare you that and just let you know that if we do so, we realize that to maximize the likelihood that our features map to the correct labels, it is enough to minimize a relatively simple equation. This equation is the loss function $L = \frac{1}{2}(h_\theta(x) - y)^2$.

In essence, an artificial intelligence algorithm is simply a good probability estimator, and it uses probability theory to develop the plan for estimating an appropriate answer to our questions. Once we have established our plan for calculating the probability of our results being accurate, we can use calculus and gradient descent to adjust the model weights so that we can correctly follow our plan. This is important to understand if we want to stop thinking of artificial intelligence systems as some metaphysical entities that, for all we know, "might be planning to enslave us." Understanding the inner workings of these algorithms reveals that the foundation of AI is probability—and, therefore, also uncertainty. This means that while we can rely on AI systems to help accelerate much of the work we do, we must first understand their strengths and weaknesses before we can accurately estimate the benefits they represent. Even when they function well, it only means that, based on their training data, there is a high probability that their answer is correct.

LOGISTIC REGRESSION

In the previous section, we learned that linear regression is used to forecast values by finding trends in a data set. It constitutes one of the main pillars of artificial intelligence, and it is the predictive basis for many neural network architectures. The other pillar of artificial intelligence is logistic regression. Logistic regression—a classification algorithm—has been around for a long time. One of the most popular tasks for AI algorithms is to classify items into different categories. Logistic regression lets us classify items into two categories. We might have a data set containing patient information: age, blood pressure, blood sugar levels, smoking habits, drinking habits, family history of heart disease, and many other possible indicators of heart disease. Our task in this case might be to classify the patients in the data set as likely to suffer from heart disease or not. Another example where logistic regression might be used is in the hospitality industry. Suppose the owner of a resort wants to identify the guests that are most likely to return to the resort based on some data about the guest: length of stay at the resort, amenities visited, money spent in the resort, age, and so on. Using this information, the resort might want to tailor their services to ensuring that those guests indeed return, or they might want to reach out to the guests that the algorithm predicts are unlikely to return, to see if they can change their guests' minds.

In computer vision, logistic regression or softmax regression (a generalized form of logistic regression with multiple categories, as first mentioned in chapter 1) is used to classify images into different categories: cats, dogs, cars, trees, and so forth. In the last layers of almost all classification neural networks, there is either logistic or softmax regression. We do not need a neural network to perform logistic regression. We can use logistic regression directly on a data set. Neural networks are used to reduce the size of the input vector into something that is more manageable for logistic regression to handle.

Recall from figure 3.1 (way back at the opening of the chapter) that the reduced form of the input is what we call the latent vector. In rare cases where the nature of the data is simple, and the samples are not described by many features so that the input vectors are small, we can skip the neural network and perform logistic regression directly.

Similar to linear regression, with logistic regression our task is to find a line. In this case, however, instead of finding a best-fit line through the data, our task is to find a line that separates our data set into two categories: 0 and 1 (or cats vs. dogs, or blue vs. orange) (fig. 3.12). The line serves as a classification model because when we get a new data point, we can predict its category based on which side of the line it falls on. To train a logistic regression model, we need a training data set with labeled samples similar to what we did in the linear regression case. A data set that is used for classification has categorical labels, whereas linear regression uses numerical labels. Numerical labels represent quantitative values that a model needs to predict (e.g., the price of a house). Categorical labels represent categories that a model needs to predict, for example, 0 or 1 to denote if a patient has heart disease or not, or 0, 1, 2, 3 for cars, cats, dogs, trees.

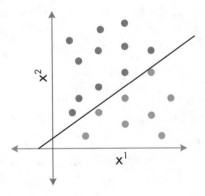

Figure 3.12 A logistic regression model separating two classes of samples (*blue dots* and *orange dots*) into different categories. In this figure, x_1 and x_2 are simply the two features describing our 2D data.

At this stage of our discussion, we will stick to binary classification problems as it makes the conversation easier, but multiclass classification follows the same concept. When we were working with linear regression, our goal was to find a line that best fit the data, and the line itself became the model. To find the line, we started with the equation of a line, $y = \theta_0 + \theta_1 x$, and our training process involved discovering our θs. Once we discovered the values of our θs, to predict some numerical value for a new x, all we had to do was plug x into the equation and solve for y. Simple.

With logistic regression, the concept is similar, but the equation is different. Remember that logistic regression is used for binary classification, where an output of 0 corresponds to class A and an output of 1 corresponds to class B, so the predictions of the model must be between 0 and 1 for any value of x. What we need, then, is an equation that can output a value between 0 and 1 for any given input. For this we use an equation called the *logistic function*: $\frac{1}{1 + e^{-(\theta x)}}$. The logistic regression algorithm gets its name from this function. The advantage of the logistic function is that it produces values between 0 and 1 for any values of x (fig. 3.13). To understand the concepts of classification, we don't have to dig too deeply into this equation. The equation is there simply to help us turn input values (these are the x values in the exponent of e) into a range of 0 to 1 output values. The e value, also known as Euler's number, is a mathematical constant that can be used to express the natural exponential function $f(x) = e^x$. If you are interested, googling Euler's number and learning about the natural logarithm can be a thrilling experience, but for our purposes, we can leave it at that.

For logistic regression, our goal is still to find the parameters of the model—the θs. We want to find θs such that our predictions for y are close to 0 for inputs whose labels are 0 and close to 1 for inputs whose labels are 1. If we think of our goal as still finding a line, in this case our line won't be a best-fit line. The points along this line are not the model's

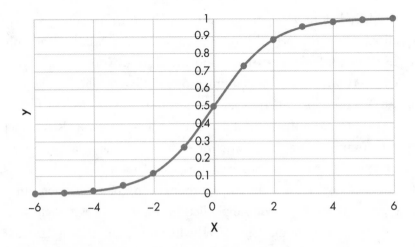

Figure 3.13 Possible values of y ranging between 0 and 1 for any input x for the logistic function, $\frac{1}{1 + e^{-(x)}}$.

predictions, as in the case of linear regression. Instead, with logistic regression, the line represents the decision boundary that best separates the two classes. For example, consider a data set of images of cats and dogs. If we can interpret each image as a vector in some space (images are made up of pixels, and each pixel has a value, so we could consider the pixel array of an image as a vector in some multidimensional hyperspace), and if we were to plot the images of cats and dogs in a graph, we would see that the images of cats are somewhat bunched close together, and the images of dogs are also bunched close together, and for the most part, the two groups are far away from each other. This might sound familiar as we have discussed vectors as lines traveling through space in some direction, where vectors pointing in a similar direction describe similar data. At some point in this space between the two groups runs an invisible line that separates the two groups. The farther away you go from the line in one direction, the more doglike the images look; the farther away you go from the line in the opposite direction, the more catlike the images look. This is the line we are trying to find with logistic regression. We are trying to find the line that best separates a group of samples into two groups.

Initially, before we begin training the algorithm, our θs are randomly initialized. We can expect this to produce a line that will be randomly placed in our sample hyperspace, and therefore it won't be classifying our data very well at all. The training process is like the linear regression process. We use a loss function, and the loss function tells us how far our predictions are from the ground truth label for that sample. Remember that our training data set consists of data samples and labels. So we would typically have thousands of images of cats labeled 1 and thousands of images of dogs labeled 0. Early in the training process, when our θ values are not well adjusted, the predictions for a cat image might be a value that's closer to 0, maybe 0.27. The loss function helps us calculate how far 0.27 is from the desired label, 1. As we discussed in the linear regression section, choosing the right loss function for a problem is the most important step in the learning process. Typically, the loss function is related to a certain probability distribution we want to model. The loss function for the logistic regression algorithm is called the *logistic loss*, or, sometimes, the *binary cross entropy loss*, and it is written as follows:

$$L = -y \log\big(h_\theta(x)\big) - (1 - y) \log\big(h_\theta(x)\big)$$

Again, there is no reason to panic. The equation is, in fact, quite simple. We do not have to understand all of it. Here is all we need to know: y is the ground truth label for a sample, and $h_\theta(x)$ is the model's prediction for a sample with features x. Recall that we already discussed this when interpreting the regression loss; $h_\theta(x)$ is simply what we call our model. It is basically our hypothesis for what the input features map to. We don't have to look too deeply into the mathematics behind that equation; all we need to do is understand the intuition. The only element in the function we might not have seen yet is the logarithmic function log( ), and thankfully calculators have a button for this—whew!

Let's see with an example how this function might be used. Suppose we have an image of a cat with a ground truth label $y = 1$. What happens if the model predicts 1? In this case, $y = 1$, and $h_\theta(x) = 1$.

$$L = -1 \log(1) - (1 - 1) \log(1) = 0$$

If our model predicts 1, then our calculated loss becomes 0. This makes sense because a prediction of 1 matches the ground truth label in our example, so we would want the loss to be 0. A loss of 0 means that in this training iteration, we do not have to adjust our model parameters. After all, our prediction was exactly right, so why would we want to adjust our weights? But what would happen if the model predicted 0 instead of 1? In this case $y = 1$, and $h_\theta(x) = 0$.

$$L = -1 \log(0.0001) - (1 - 1) \log(0.0001) = 4$$

If our model predicted a 0, our loss would be 4.[17] In this case, our loss is higher than 0. This also makes sense because the model predicted 0 where the ground truth was 1, so we want our loss to signify that our model has made a mistake and requires correction. How do we adjust our model? Similar to how we responded to a like situation in linear regression, we use the gradient descent algorithm to find the values of θ that produce the smallest loss for the entire data set. As a reminder, the θ parameters are updated for each training iteration as follows:

$$\theta = \theta - \alpha \frac{\partial L}{\partial \theta}$$

The new value of θ is equal to the current value of θ minus the gradient (or rate of change of the loss with respect to θ) times a learning rate parameter. Again, to explain this more simply, we use calculus to find

---

17. Mathematically, $\log(0)$ is undefined, so if our model $h_\theta(x)$ predicts a value of 0, our training algorithm must add a small quantity to keep the value close to 0, just not 0. Therefore, $h_\theta(x)$ is set to 0.0001 in our example.

the rate of change of the loss with respect to the current value of θ. This means finding out how much the loss changes if we change θ a little bit. The rate of change tells us whether increasing θ increases or decreases the loss. We want to change θ in the direction that decreases the loss, as this leads to accurate predictions.

In the linear regression example, we saw α in the gradient descent equation and deferred on its explanation, as we wanted to focus on describing the most important aspects of the algorithm. Here we see it again. The parameter α is what we call the *learning rate parameter*. It is a tunable parameter that specifies how much we should change θ in the direction that minimizes the loss. Finding the right value for α is very important. If α is too large, as we are traversing our loss surface (recall our thought experiment where we imagined the loss surface as some topological terrain with mountains and valleys), we might take too large of a step. A very large step might be akin to jumping from one mountain peak to another. These jumps risk our overshooting the valleys (the loss minima), so we end up in another part of the loss surface where the loss is large. Large jumps can lead to algorithms that never learn (or never converge). Conversely, we do not want our descent into the valleys to be too timid. If the steps we take in our descent are too small, it may take too long to converge on the loss minima. Imagine walking down a mountain by taking baby steps. Unfortunately, finding the right learning rate value can be more of an art than a science. It is called a tunable parameter, or hyperparameter, because we must choose the value ourselves. The training algorithm does not attempt to find it. In practice, we run multiple experiments or training cycles using different learning rate values. By running multiple experiments, we can find the right step size that gets us to convergence in a practical amount of time.

Let's do a quick recap on logistic regression. Logistic regression is a classification algorithm that helps us find a line that divides a group of data samples into two classes: the class to the right of the line and the class to the left of the line. To train the algorithm, we minimize the following loss function using gradient descent, $L = -y \log(h_\theta(x)) - (1 - y) \log(h_\theta(x))$. Once the algorithm is trained, to classify a data sample, we ask the following question: $\frac{1}{1 + e^{-(\theta x)}}$. We present the logistic equation with our sample's features x and the learned parameters θ and calculate our output. If the result is close to 1, our prediction is class 1. If the result is close to 0, our prediction is class 0. It is that simple! But why does it work?

When we discussed linear regression, we described the probabilistic interpretation of the learning process and explained that, in essence, minimizing the loss function was akin to maximizing the likelihood that the sample features mapped to the result we wanted the model to predict. And this was made possible by relating the loss function to a probability distribution, because probability distributions help us calculate the probability of making certain observations. So, if you have a problem and you don't know the solution to that problem but you can calculate the probability that a solution is correct, you might not have the final solution yet, but at least you have a path forward. In fact, you have more than that. You have the ability to measure success in each step that you take toward a working solution. This is the power of probability distributions!

In the linear regression case, we chose the normal distribution because we assumed that the noise scattering the data samples about our predictor line was distributed according to the normal distribution. We said that we can think of minimizing the loss function as being similar to discovering the parameters of the normal distribution describing our data, and once we learned those parameters, we could sample from that distribution to guide our best-fit line. It turns out that for logistic regression, the problem we must solve is very similar to our problem in linear regression. We also have a loss function, and we must minimize the loss function to update the model parameters following the same gradient descent algorithm. So what is the difference between linear regression and logistic regression? If the training process for both is the same gradient descent algorithm, how come in one case we get a best-fit line and in the other case we get a line separator? The difference, of course, is the loss function.

The logistic loss is different from the linear regression loss. This is because logistic regression is modeling a different probability distribution than linear regression. For logistic regression, the probability distribution we are trying to model is the Bernoulli distribution. The Bernoulli distribution describes situations where we are performing a single trial (e.g., tossing a coin or predicting a cat or a dog) and the trial can result in one of two possible outcomes: success or failure. The Bernoulli distribution helps us predict the likelihood of either of those outcomes occurring. Let's see how our loss function for logistic regression relates to the Bernoulli distribution.

PROBABILISTIC INTERPRETATION OF LOGISTIC REGRESSION

First, we will try to develop an intuition for how this sort of distribution can be described. Since we are discussing cases that can only have one of two possible outcomes—success or failure—we can describe the probabilities as follows. Let's say that the probability of success for a given experiment equals some probability p. The probability of failure, then, must be 1 minus the probability of success. More formally this can be described as P(success) = p, and P(failure) = $1 - p$. As an example, we can consider what happens when we are tossing a fair coin. First, we can assign success to flipping heads in a single trial and failure to flipping tails. The probability of the coin landing heads would be 50 percent, so P(success) = 0.5. The probability of failure, then, is P(failure) = $1 - 0.5 = 0.5$. This makes sense because tossing a coin results in 50 percent probability for both heads and tails.

Note that the probability of an event occurring does not have to be 50 percent. Here is an example where the probabilities of success or failure are not equal. Suppose we are describing the chances of a particular Toronto-bound train being delayed. We know from experience that the train is punctual 95 percent of the time. We can describe the probability of the train being on time as P(success) = 0.95 and the probability of a delay as P(failure) = 1 − 0.95 = 0.05.

To make our lives easier, we can also combine the P(success) and P(failure) expressions into a single equation: $P(x) = p^x(1 - p)^{1-x}$. Let's use this equation on our train example to show that it is truly combining both P(success) and P(failure). We can define $x = 1$ as success and $x = 0$ as failure. We can say that the probability of the train being on time (i.e., success) is $P(1) = p^1(1 - p)^{1-1}$. Any number elevated to the power of 0 is 1, so $P(1) = p^1(1 - p)^0 = p$. This is exactly P(success) = p, which is 95 percent probability of our train being on time. Now let's check on the failure case. $P(0) = p^0(1 - p)^{1-0} = 1 - p$. This is exactly P(failure) = $1 - p$, which is 1 − 0.95 = 5 percent probability of the train being delayed. We have shown that $P(x) = p^x(1 - p)^{1-x}$ combines the probabilities of success and failure for a Bernoulli event.

You can now also start to see some similarities between $P(x) = p^x(1 - p)^{1-x}$ and our loss function $-y\log(h_\theta(x)) - (1 - y)\log(h_\theta(x))$. We are getting closer to explaining where our loss function comes from. If you can already see that the loss function is a restatement of the p elements in the Bernoulli equation in terms of our model and data labels, then you are done. You should now understand the origins of the logistic regression loss function. If you don't see it yet, don't worry. We will explain it next. Let's see one more example.

Suppose there is a place somewhere on earth where it rains once a month, and every day has an equal chance of rain. If we pick any day of the month, the probability that it will rain on that day is 1/30, that is, P = 1/30. Since we are trying to predict rain, let's call success 1 and failure 0 so that P(1) = 1/30. The probability that it won't rain on that same day is P(0) = 1 − 1/30.

Now suppose that we are not lucky enough to know the probability that it will rain on any given day of the month. In other words, no one has told us that the probability is 1/30. Instead, we are given a data set containing the days that it rained in every month for the last ten years, and we are asked to create a model that can predict whether it's going to rain next Thursday. What we want is to build a model that can learn that probability on its own. In this case, we can treat the data as being distributed following the Bernoulli distribution. We can do this because the Bernoulli distribution describes events that only have one of two outcomes, and for each day, it either rained or it didn't. We can then build a model that tries to find the parameters that describe the Bernoulli distribution for this data set. And once we do that, we can predict the likelihood that it will rain next Thursday. So how do we go about doing this?

We start with the Bernoulli equation $P(x) = p^x(1-p)^{1-x}$. Remember that we don't know the probability of when it will rain; all we have is a data set. We can rewrite the Bernoulli equation in terms of our model and training process. $P(x)$ really means the probability that for some sample x, the model predicts the correct label y. So $P(x)$ is $P(y|x; \theta)$. Recall that we can interpret this statement as the probability that sample x maps to label y in a model with parameters θ. The right side of the Bernoulli equation can also be rewritten in terms of our model. We know p is simply the probability of some event happening, which in terms of our model refers to the model predictions. Therefore $P(x) = p^x(1-p)^{1-x}$ can be rewritten $P(y|x; \theta) = h_\theta(x)^y(1 - h_\theta(x))^{1-y}$. These are all terms we should be quite familiar with by this time. Now we need to train our model to maximize the likelihood that our predictions $h_\theta(x)$ produce the correct output for all samples x. If we devote the next three pages to algebra and develop all the necessary steps—I will spare you—we can show that maximizing the likelihood of our predictions being accurate is the same as minimizing the output of the following equation: $-y \log(h_\theta(x)) - (1 - y) \log(h_\theta(x))$. Interestingly, this just happens to be our loss function.

What have we achieved in this process? We have learned that whether we are performing linear regression or logistic regression, the training process of an AI model is very similar. All that changes is the loss function. The loss function is typically related to a probability distribution, and probability distributions are the secret sauce of artificial intelligence. There is nothing else that drives an AI model's predictions. There is no metaphysical "intelligence." There are no premonitions or revelations. When we lift the veil and peek into every single operation that is happening inside our AI models, all we see is math. The reason these models have any right to work at all is not because of luck or because, by some dark magic, a mess of calculations suddenly learns to recognize patterns. The reason they work is because the models learn the probability that each sample belongs to each of the output categories. And the models can learn those probabilities because of what mathematicians and statisticians have done over centuries to describe probability distributions.

Probability distributions are great at describing events in the world, and it turns out that many events that appear quite distinct can still be described by the same probability distribution; all that changes are the values of the parameters for the distribution. For example, we can use the normal distribution to model population height, but we can also use it to model the birth weight of newborn babies. The parameters of the distribution—the mean and variance—will be different for the height and the birth weight distributions; the parameters of the distribution will also be different for each country. But knowing that we can use the normal distribution to model those processes means that once we learn the mean and variance for the distribution, we don't need to know exactly the birth weight of every baby in a given country; we can approximate it. If I find anything truly inspiring and ingenious about AI, it is the realization that we can use these *world models* that are probability distributions as the predictive engines for mathematical approximation functions. We call these mathematical approximation functions *AI models*.

○ ● ○

We now find ourselves at the end of a math-intensive chapter. Thank you for reading thus far. My goal with this book has been to describe how our most successful artificial intelligence algorithms work. The mathematics behind these models can seem daunting at first, but I hope to have shown that, once we understand the intuition behind the calculations, understanding how and why these systems work is not so difficult. In this chapter, we described in detail the functional aspects of neural networks—the bit that makes them work. In the previous chapters, we described neural networks as funnels that compress information from the input layer into a latent vector representation. This latent vector representation is a compression of the input. In other words, it is a distilled version of the input, where the neural network has learned to discard bits of information that are unnecessary. This latent vector representation is used in the last layer of the neural network to perform either classification or some form of linear regression. In this chapter, we also described how linear regression and logistic regression—the most common form of classification—work. We showed that if the data we are trying to analyze is simple and the dimensionality of the feature vectors is low enough (i.e., if each data sample is described by a handful of features), we can use linear or logistic regression directly without requiring a neural network. We use neural networks when the dimensionality of our data sample is so large, with so many features describing the data samples, that we require some form of compression first.

You should now have a good foundation for understanding the intuition behind most state-of-the-art artificial intelligence algorithms. Is it a technical understanding that enables you to go off and do research or write your own AI algorithms? No, but if that's your goal, then the last three chapters should have illuminated the path ahead and shown the direction to follow to expand on your technical knowledge. Now you know what gradient descent means and what probability distributions and loss functions are. You can continue to dive into these topics and become an engineer or scientist if you so wish. If that's not your goal, however, if your goal is simply to understand how neural networks function so that you can have an intelligent conversation and formulate your own opinions about the ethics of AI, then the last three chapters should have given you the tools to do exactly that.

In chapter 1, we learned about the history of the neural network and its basic architecture. In chapter 2, we learned about computer vision and a popular architecture called the CNN, which is great for image processing. We learned that we can combine a CNN with a fully connected network—the architecture from chapter 1—and build good image-classification algorithms that power many of today's computer-vision solutions. In this chapter, we learned what makes all that possible: probability distributions, calculus, and optimization through gradient descent. Now we know what neural networks are: mathematical approximation functions. And we know what they are not: semiconscious creatures salivating at the chance to break loose from virtual shackles. In the next chapter, we discuss what we can do with AI and, now that we know how it works, what our ethical responsibilities are.

Is there anything to fear about AI, or is it all fair game?

4

INTELLIGENT
DISCOURSE

Is the algorithm plotting against us? That's the question that started it all.

As early as in the introduction, we revealed that our current neural networks are not conscious systems, and whether they are truly *intelligent* depends greatly on our definition of intelligence. Over the past three chapters, we learned how neural networks operate. Now we understand the structures and calculations that enable them to function in surprising and remarkable—but not magical or sentient—ways. So, are Alexa and Siri conspiring to take over Earth? Maybe. But if they are, it's not personal. It's just gradients. From this perspective, even if you consider AI systems the bogeyman, at least now you know which bogeyman to fight.

We have arrived at the point where we need to evaluate what we have accomplished and consider how, in each of our individual capacities, we might best apply what we have learned. This is not a technical book. The purpose of the book has not been to teach engineers how to implement neural networks. The purpose of the book is to give everyone interested in recent developments in AI (and now we know the extent to which any of this is recent) an understanding of how these systems work.

The burden of knowledge is twofold. First, we must devote considerable time and energy to digest a new concept. We need to build new conceptual visualizations and analogies to internalize the information we receive so that it is easily accessible and applicable to our own lives. Think about anything that you have learned and have been able to retain and recall at opportune times. The process of learning those things was more complex than simply having to memorize a string of facts. You had to invest significant resources to create mental images and models to relate with the problems and the situations, and you had to imagine a path from the problems to the solutions. This energy was well invested because knowledge empowers us to change the quality of our lives. Think about how far we have come since the first dark cave was suddenly illuminated by a controlled fire. But knowledge also makes us responsible not only for any action we take in the exploitation of our newfound power but also for our lackadaisical attitude toward reigning in unintended consequences.

Burning fuel literally ignited the Industrial Revolution. It empowered us to build bigger and more powerful machines that, arguably, improved our lives. But we now know the consequences of burning fuel, for the planet and for ourselves. It is up to us to decide what we do with that knowledge. We can use our voice to influence policy, or we can close our eyes and pretend all is well. Either way, once you know, absolution is forfeited.

Artificial intelligence, and the current algorithmic revolution, is in principle no different from any of the previous technological revolutions our species has endured. It is argued that the first technological revolution, the agricultural revolution, gave us a certain level of security (food security) at the expense of freedom. When we were simple hunter-gatherers, we were able to roam the land freely, moving to a different place with each season or according to whatever needs arose. The agricultural revolution and the domestication of wheat led to farming. But farming meant that we could no longer move as freely about the land as we once did. The new crops were feeble and needy, requiring our constant attention. We were now bound to a small piece of land for the rest of our lives.

Each new advancement and each piece of technology that we create will always have positives and negatives, and artificial intelligence is no different. It is difficult to balance the books on transformational revolutions to know whether the net effect was good or bad. The agricultural revolution forced groups of people to work together, which created close communities. Large groups of people living together in proximity led to a decrease in hygiene and the rise of new diseases. For a very long period, our life expectancy shortened compared to when we were hunter-gatherers. It was not until the advent of modern medicine, antibiotics, and vaccines that things began to improve. But although living in communities had its downsides, it also led to an increase in communication. We had more people working together to solve problems. Working together and growing our communities into towns and cities forced us to create new technologies and made us the most powerful species in the history of the planet. Was it worth it? Well, we are the most powerful specifies to ever walk this planet, but it is also very possible that we will be the shortest-lived species in the history of the planet. It is difficult to calculate the worth of the trade-off. Those lucky enough to have lived in prosperous periods for our species might say that it was worth it, but the generations living in the dusk of our civilization and those who will miss out altogether might have a different view.

I don't know whether we are truly incapable of containing the technologies we create or whether our desire to explore all aspects of technology—the good and the bad—is just too great for us to control. For example, once the power of the atom was understood, was the nuclear bomb inevitable? Could we have used nuclear power simply as a source of clean (cleaner?) energy? Or were we immediately cursed with the need to create an exit button (for a generations-long, species-wide existential crisis) and send ourselves straight to the halls of extinction? Clearly a choice existed where we did not have to invent a nuclear bomb, but as a species, it seems that we are not mature enough to have made that choice.

I think we now find ourselves at a similar point in history. We live in the Age of Data. We generate more data every second today than our species generated (or at least recorded) in our entire history. We have unlocked new powers with the discovery of new technologies (e.g., artificial neural networks and the use of GPUs and their massively parallel computing engines) and the implementation of tools that help us make sense of the massive volumes of data we are generating. The danger is in blinding ourselves to the limitations of our newfound powers and automating large parts of our lives using technologies we don't completely understand while completely disregarding the limitations of those same technologies.

I began this section by saying that artificial intelligence is no different from any other technological revolution we have ignited. But that's not quite true. In some ways, it is very different. With every other technological revolution, we have made both good and bad choices. But throughout, the choices have been ours. This time, we are potentially looking at a different set of problems. We can accept the wide adoption of AI-driven automation in every corner of our lives, and that will be our choice, but that's the only choice we get to make. Once adopted, the choices that will shape our future will not be our own.

THE CURSE OF INTELLIGENCE

In the first part of the book, we were gathering tools. Just as the early humans sat at the mouth of a cave shaping a piece of obsidian or flint to make a knife, so were we forming our understanding of artificial intelligence to make a set of tools that will help us establish a conversation. It would be impossible to discuss artificial intelligence and develop an opinion on its benefits and dangers, as well as our responsibility toward it, without first defining what artificial intelligence is. In this book, we have centered our discussions of artificial intelligence on the neural network. Neural networks are not the only algorithms capable of making sense of data, but in many areas of research, they certainly represent the current state of the art.

Neural networks also possess a mystical aura just because they sound biological. It's hard to say "neural network" (even if we say "*artificial neural network*") and not conjure up an image of a thinking brain. If we want to consider artificial intelligence and evaluate its potential impact, we need to define what we mean by artificial intelligence. The term *artificial intelligence* suffers from the same misconceptions as the term *artificial neural network*. The problem is that the words *neural* and *intelligence* are overloaded with meaning. We know what intelligence means. A person who is said to be intelligent possesses a wealth of knowledge and can apply that knowledge to solve complex problems. But even this is a narrow definition of what we think of when we think of intelligence. When we think of intelligence, we also think of awareness and experience. Most of the time when we consider intelligence, our definition is broader than simply "having knowledge and applying it." Instead, intelligence as a general concept combines both being aware of the implications of the problem we are trying to solve and being able to draw from experience and apply knowledge in a completely new way—in other words, a crucial aspect of human intelligence is our species' ability to *innovate*.

Consider a child and his babysitter playing at the beach. The child digs a hole in the sand and wants to fill it with water. He realizes that he needs a bucket to bring water from the sea's edge to the hole. He asks his sitter to bring him the bucket from the bag of beach toys. She goes to look for the bucket and realizes they left it at home. She sees a tantrum on the horizon and feels a stab of frustration at the realization that the beach day is about to be ruined. How could they have left such an important piece of equipment behind! Then she notices the small plastic container his mother used to pack them a snack. It's a square container, but it's deep enough to hold a good amount of water. She empties the snacks into her purse, and suddenly the crisis is averted. They might have left their bucket behind, but she found a workable solution, and the water hole is filled in no time.

Let's analyze this situation and see if we can learn something from it. The child knew that to carry water from one place to another, he could use a special tool called a bucket. Let's replace the child with a robot. If we ask a robot to get us a bucket to fill a hole with water, and if the robot can identify a bucket among a pile of toys and bring us the bucket, we might think of that as intelligence. The robot could understand our instructions: "Get a bucket." It could identify an object; that is, it could apply its knowledge of what a bucket is to identify it among a group of toys. But what about the babysitter who couldn't find the bucket and used a storage container instead? The process of identifying the container as a bucket replacement required awareness of the problem. In this case, the sitter was able to abstract from the literal definition of the bucket and consider its purpose. Ultimately, she didn't need a bucket. She needed a method to get water from the sea to the hole. Once she made this realization, she was free to explore beyond the bucket. She drew from past experiences moving water around. She knew that other containers can also be filled with water. A cup can be filled with water. A bottle can be filled with water. And a storage container can be filled with water!

Can our robot do this? The robot we have described, which is only capable of understanding our command at a very literal level and has a vision algorithm that lets it identify the object we requested among a group of objects, would not be able to find a bucket replacement. It can identify a bucket and retrieve it, but it is not capable of innovating when the bucket is missing. To be fair, this is a simple example, and there are more sophisticated algorithms we can use that have a good chance of helping a robot come up with a bucket replacement to haul water. But the point I am trying to make is that there is a difference between the narrow definition of intelligence, where information is applied directly to problems, and the broad definition we often use when discussing a general intelligence that's based on awareness and experience and that involves a healthy amount of creativity. This is the curse of the term *intelligence* when applied to the field of AI. To the uninitiated, the word immediately conjures up and attributes vast powers to artificial systems. But as we have seen, there is still a large gap between human-level intelligence and our best efforts at artificial intelligence.

APPROXIMATING A GENERAL INTELLIGENCE

There is a class of algorithms we haven't discussed, and in their functioning, these *reinforcement learning* algorithms come close to innovation. Some researchers believe they may be the best path we have to an artificial general intelligence. It's useful to discuss reinforcement learning briefly because such algorithms have generated the most media attention in recent years by enabling a machine to beat human players at the games of chess and go. But we can see that even these systems are still too constrained to match human-level intelligence or anything resembling consciousness, though they have demonstrated that our species will never again dominate chess.

Reinforcement learning (RL) algorithms are quite different from the artificial intelligence algorithms we have examined thus far, which are known as *supervised learning* algorithms. As we saw in the previous chapters, these algorithms begin their training with a data set that has been fully labeled by a human. These algorithms learn to map the sample inputs and their labels. In RL, there is no data set, and there are no labels. In the RL framework, there is an *agent*, and there is an *environment* where the agent *interacts*. For our purposes, the agent is the learning entity. The agent makes a change to the environment and evaluates its new position in the environment. The learning process encourages the agent to maximize a specific reward. For example, consider an RL framework built around the game *Tetris*. In this example, the agent is the player, and the environment is the *Tetris* playing field.

We want the agent to learn to play the game *Tetris*. The agent can change the environment by providing a set of inputs to the game: move the incoming piece left, move the piece right, change its orientation. What's impressive about reinforcement learning is that it requires no interactions or explanations from a human. The agent initially doesn't know anything about *Tetris* and doesn't even know how to win or lose the game. The agent simply knows that there are three types of moves it can make for every new piece that shows up: left, right, and change orientation. The only measure of success the agent has is a reward system. In this case, the reward is maximizing the game score. The agent proceeds to play the game by making moves and monitoring how its moves are affecting the score. Moves that increase the score receive a positive reward, and moves that worsen the agent's position—for example, by filling up the field and losing the game—receive a negative reward. After several hundred (or thousand) rounds of play, the agent learns strategies for keeping the rows of bricks low and pushing the score higher. It has successfully learned to play *Tetris*.

If this sounds like an ingenious learning mechanism, it certainly is. But it is easy to oversell its implications and current potential for a general intelligence. The truth is that reinforcement learning is one of the most complex classes of artificial intelligence algorithms we have. There is a high degree of freedom in framing the problem in a way that maximizing a reward can lead to successful solutions. For example, consider that some good moves are only good in retrospect. In other words, sometimes a great move does not immediately increase the score, but it does set up the playing field in such a way that five moves later it will contribute to a hefty score increase. Considering how to frame the problem of *Tetris* (or the environment in general) in a reward-based mechanism where the agent's position can be evaluated at every step is where reinforcement learning becomes very difficult to master. And it is one of the main reasons why it isn't yet the state-of-the-art solution for all artificial intelligence problems.

It is difficult to know whether RL will eventually get there or whether another algorithm will surface that can adapt to the environment and learn without interactions from humans, all while being a less cumbersome framework to deal with. RL has, however, provided really surprising solutions to questions that at one point were considered impossible for an AI system to solve. In 2017, DeepMind (a subsidiary of Google) made headlines when its AlphaGo algorithm beat Ke Jie at a game of go. Ke Jie was, at the time, the number one ranked player in the world, and go was considered the most difficult game to conquer by an artificial intelligence, with many more possible combinations of moves than even chess. The next version of the algorithm, AlphaZero, is currently considered the top player in the world for both go and chess, and many believe that at this point no human could ever beat it.

Advancements like these in AI research are impressive and contribute a building block to the process of one day developing a general artificial intelligence, but they also give the public a false sense of progress. Chess, being a difficult game for most people to master, has a special place in our minds as a gatekeeper for intelligence. If you can play chess well, you are automatically considered "smart." When AlphaZero is discussed in the media, it is presented as more intelligent than a human (because it is a remarkable algorithm that can beat the best human players at both chess and go) and maybe weeks away from formulating a plan for world domination.

When we compare *Tetris* to chess, the difference in complexity between the two games is so pronounced that it might suggest AlphaZero is a few evolutionary steps beyond the much simpler RL algorithm for beating *Tetris*. Since most of us can fare pretty well in a game of *Tetris*, we might not be all that impressed by an algorithm that can dominate the game, even if it far outperforms us. And if it can't beat *Tetris* and we can, then it certainly can't do other things that we can do well. But AlphaZero can beat any of us at go and chess (and *Tetris*, were it given the chance); therefore, we automatically worry about what else it can do better than us.

What is interesting about AlphaZero is that it is fundamentally no different from the reinforcement learning algorithm that beats *Tetris*. The framework is similar. There is an environment and an agent. The agent makes changes to the environment (moves in the playing field) and evaluates its position relative to a reward system. The AlphaZero algorithm can make moves and discover plays that human grand masters consider "brilliant" and "creative," but AlphaZero doesn't know it's being clever. It's not chuckling to itself and exclaiming, "Ha ha, I got him with this one!" The real magic in AlphaZero comes from the ability of the researchers who created it to frame the problem of chess in a way where every move can be evaluated relative to a reward that must be maximized. This is especially difficult with chess, where great moves are notorious for having delayed benefits. It's very difficult to evaluate the current position in a game of chess because advantages can be very subtle. For example, a player may make a move that costs him a high-value piece. A shortsighted reward system may penalize this as a bad move, having lost a valuable piece. But the move may have sacrificed the piece to open the space for a particular attack ten moves later.

This means that an algorithm that is going to learn to play chess at a grand-master level must learn to evaluate the positions with a long view of the game. This makes developing a reward system that evaluates every move extremely difficult. And this is what is truly remarkable about AlphaZero. The irony is that the genius of AlphaZero isn't AlphaZero. It's the human researchers who designed it! The impressive feat is that a group of people figured out a way to set up a reward system and a score-tracking system that penalizes bad moves and rewards good moves in chess—where the system itself discovers what is "good" and what is "bad" simply by playing and evaluating its position. Fundamentally, all the algorithm is doing is learning a distribution of probabilities for each move it can make for the current position, where a high probability signifies a potentially good outcome.

That's right—like the other algorithms we saw earlier, AlphaZero is also learning a set of probabilities. It doesn't even know it's playing chess (or go). Just because it can beat a chess grand master does not mean that it can also drive a car or write a novel or even help a robot climb a set of stairs. This is very important to grasp because it demystifies these algorithms and helps us evaluate progress against reality. We might certainly be able to use the concept of reinforcement learning to create a framework that can drive a car, help a robot walk up a flight of steps, and maybe even write that novel. But for each of these use cases, someone will have to frame the problem in terms of rewards and develop a method for evaluating each move or change. And these reward systems and evaluation metrics will be different for driving a car and writing a novel; an algorithm that can learn to drive a car by playing with the environment will not necessarily learn to write a novel using the same reward system.

This means we do not yet have a method with enough plasticity to create a single algorithm that can learn to do everything we can do—in other words, that has a general intelligence. Research is still ongoing in this area, but we are not there yet. It may turn out that a single algorithm cannot generalize to learn to perform all the different tasks humans can do. It may be that we need a set of algorithms working together, each capable of performing a set of specific tasks in different domains. But currently, what we have is a set of algorithms that perform quite well at very specific and well-defined tasks.

RESPONSIBLE ARTIFICIAL INTELLIGENCE

We have artificial neural network algorithms that can classify images. These algorithms, as we saw, are quite successful in the field of computer vision. We have algorithms that process natural languages and can write prose that is so well written and coherent that it is indistinguishable from something produced by a human. But all these algorithms have limitations in that their performance is directly tied to the data they were trained to model. An algorithm that was trained to distinguish between oranges and apples will not be able to recognize certain diseases in chest X-rays, and more importantly, the process of retraining these algorithms to learn from a new domain isn't trivial. I don't think we know exactly how our brain works and whether there is a single all-encompassing biological algorithm or a collection of them highly tuned to specific tasks. Some scientists believe it's a bit of both. But although our brains may use different algorithms to solve different classes of problems, our brains can reconfigure themselves to learn these algorithms and create the necessary neural connections that let a person navigate a busy intersection or get a medical degree. Our progress in artificial intelligence has been in creating algorithms that can achieve remarkable results with speech recognition, others that can learn to classify healthy and diseased tissue from pathology images, and others still that can visually navigate busy intersections (with a few constraints, as we discuss later in the chapter), but we do not have a general *machine* that can produce the neural connections, the framework, and the necessary training for the system to learn and grow its capability, automatically, across all the domains in which humans interact.

So far, I have been trying to establish a position on the dangers of AI with respect to current technological progress and gauge how close we are to the dangers of a self-aware AI that is more intelligent and capable than humans in every aspect of life. My view is that our current algorithms are still too primitive, frail, and naive to generalize well across their own domains, let alone advanced enough to overthrow us and form their own government.

So if these algorithms can't take over the world yet, does that mean we are in the clear? Does it mean there is nothing to worry about, and can we proceed happily with the general adoption of AI in as many areas of our lives as possible? In other words, is the latent—and, at times, overt—fear of AI a justified reaction, or is it all a misunderstanding? As I detail below, my view is that there is certainly a need for discussing some very concerning trends in AI; at the same time, the proliferation of misconceptions has thwarted our ability to approach these pertinent issues, to say nothing of getting a handle on them. We are focusing too much on fantastic fears when there are more immediate problems to address. Our AI systems might be primitive, and far from becoming self-aware, but we must still employ them in responsible ways, or we will expose ourselves to terrible consequences that will be difficult to reverse.

I'd like to focus the following conversation about our fear of AI on immediate concerns rather than future concerns. Future concerns involve attractive concepts that are more closely aligned with science fiction. These are concerns around self-aware or conscious systems that can actively learn to work against us and overpower us and, indeed, topple our reign in the hierarchy of species. No wonder these concerns garner most of the attention from the media. We will briefly discuss these fears and contrast them with the current progress in AI, but our focus will be the immediate concerns. In other words, our discussions centers on the dangers that AI systems pose today, which for some reason do not get much attention. These concerns are related to how we use, and overuse, the simpler neural networks and algorithms we have discussed in this book. We do not need to wait for a self-aware blender to chase us around the kitchen. It turns out that there is plenty of harm that we can do with our current systems even if they aren't self-aware and actively trying to harm us.

THE PROBLEM OF BIAS

We touched on some of these immediate concerns in chapter 1. And bias is one such big concern. Bias exists in data, and it exists in our models, and we need to be aware of different types of biases and the

to be mean. The defect is 100 percent the fault of the designers, who somehow don't foresee that the training data set contains many biases toward the terrible.

Besides producing awful chatbots, biased data can pose a real threat to our lives. If we want to train an algorithm to assess specific health risks in individuals, then our data set should constitute a balanced cross-section of the population we want to monitor; otherwise, the predicted risks to the underrepresented samples will be highly inaccurate. In other words, we want our training data sets to contain as many samples as possible, but we also want to make sure that every category of samples is equally represented. Unbalanced data sets and statistical bias are not new, and statisticians have always had to contend with this. Sometimes bias is introduced in our data sets because of clear racism. We saw this in the Boston Housing Price data set example. Sometimes bias is unintentionally introduced due to circumstances. Consider a medical data set for cancer research. Suppose we construct a training data set by scanning images of biopsied tissue for patients who are suspected of having cancer. It turns out that most biopsied samples don't contain cancer. This is because patients who get tested for cancer don't always have cancer and because the cancer isn't always so spread out that the biopsied regions all show signs of cancer. This means that when the training data set is constructed, we should make sure to check that healthy *and* abnormal tissue are equally represented. But this still doesn't fully remove bias from the data, because even if we have thousands of samples of images of cancerous tissue and an equal number of images of healthy tissue, we may still suffer from availability bias.

How many patients were used to produce those samples? Do the samples come from ten patients or one thousand patients, and do the patients represent a good cross-section of the population? As you can see, the problem of bias is a complex one, and it isn't always immediately clear in which direction our data might be biased. So the problem of bias is neither new nor specific to AI. The problem that AI poses to biased data, which is novel, is the barrier to entry for processing the data. Consider the heart disease data set we have been discussing. We have already said it is biased toward a mostly white male population. Before AI was accessible to most people, we would have had to involve experienced statisticians to make sense of the data. In some cases, those statisticians may have purposefully or carelessly missed the limitations of the data, but in other cases, those statisticians would have flagged the biases in the data set and perhaps even have suggested ways to fix the data set. Advancements in AI and data processing, however, have made it almost too easy for just about anyone with a weekend to spare to learn how to write a few lines of code in Python and process a data set without really understanding the data. The promise of artificial intelligence algorithms is to augment our ability to process information so that tasks that used to take years for a human to perform can now take days or sometimes hours. But while our ability to manipulate data and process it has been greatly increased, the complexity of the data and our ability to really understand it have not fundamentally changed. So now we can also misunderstand data at unprecedented scales.

With this awareness, we must ask: How is this affecting us today?

ARTIFICIAL INTELLIGENCE IN THE JUDICIAL SYSTEM

We are increasingly using artificial intelligence algorithms to automate different areas of our lives, from monitoring the stock market for securities trading, to analyzing medical image pathology, to executing law enforcement and the judicial process. I'd like to spend some time discussing the use of AI in the judicial process and its implications, because to me it's one of the most dangerous use cases for AI, with the potential to bend our society toward a darker, less hopeful future.

First, let's remind ourselves of how these systems work and explore a different kind of bias. When a neural network processes information, it encodes the decision processes in the set of weights given to the neural connections. In a classification use case, the network outputs a distribution of probabilities spread out among the different outputs— in other words, spread out among the different prediction classes. Suppose we want the network to distinguish between images of apples, bananas, and oranges. The output of the network will be a distribution of probabilities that the input image is either an apple, a banana, or an orange. Notice that there is no option for "I don't know" or "None of the above." When dealing with probability distributions, the sum of the distributions must equal 1 (100 percent probability). That is, when we add the probabilities that the input image is a banana, an apple, or an orange, the sum equals 100 percent probability. This means that the network assesses that there is 100 percent probability that the input image is indeed one of the three possibilities. But note that those possibilities, those three output classes, were chosen by the neural network designer and do not represent the data set.

Basing the three output classes on what the engineer wants the neural network to detect depends on a largely incorrect assumption. It assumes that the network will only ever see images of apples, bananas, or oranges. If we are employing this neural network in a controlled environment where we guarantee that all it could ever see are either apples, bananas, or oranges, then that would be fine. Unfortunately, in practice, when we deploy these neural networks in production environments, we can't always control what they see, and we can't always guarantee they won't see something outside of what they were trained to predict. Suppose our neural network sees a bunch of grapes. The network will still output a set of probabilities that it is a banana, an apple, or an orange, and together all those probabilities will sum to 1. This is another example of bias—this time in our models. The model is biased, by design, to output only a set of possibilities, and the entire world must fit into those possibilities, and whatever doesn't fit will be jammed in nonetheless. We might naively think that an obvious solution is to add a fourth output class: the "None of the above" class. Consider, however, what this means. It means that you must train the neural network to distinguish between apples, bananas, oranges, and everything else in the universe—not a very simple proposition.

Another possible solution is to set a threshold for the probability that we accept as our network prediction. For example, we could require our prediction to be greater than 70 percent confident for acceptance. Let's assume that in the case where our network saw a bunch of grapes, the output probabilities were 33 percent banana, 33 percent apple, 34 percent orange. The probabilities are all lower than the 70 percent threshold; therefore, we determine that the network is not confident that what it saw is either a banana, an apple, or an orange. This is a very common approach to dealing with *out-of-distribution sampling*—that is, samples in the real world that exist outside of the classes of objects that the model was trained to recognize. While it is a workable solution, it gives rise to a few nuanced points that must be considered and discussed.

First, it assumes that the network will never give a high-probability prediction for an out-of-distribution sample. What if, instead of a bunch of grapes, the network had seen a cucumber. Considering that cucumbers have a shape that resembles bananas more closely than grapes do, we would expect the confidence level for the banana prediction to be higher than 33 percent. Then we need to consider what is an acceptable threshold for an accepted prediction. In low-stakes use cases like fruit selection, it may not be terribly important to be all that accurate, and if the increase in processing throughput afforded by the automation is enough to offset the cost of mistakes, the automation may still be worth it. But in cases where our future depends on these decisions and assumptions, we need to be well invested in understanding every aspect that drives the predictions.

In the United States, the use of automation algorithms has made it into the criminal courts. In 2014, the Baltimore-based nonprofit organization Pretrial Justice Institute (PJI) urged the state of New Jersey to adopt algorithms for risk assessment in setting bail. The algorithms predict the likelihood of an individual skipping court or committing another crime while awaiting trial. PJI's urging was motivated by racial inequities in the bail decision process. Their hypothesis was that a mathematical algorithm—emotionless, number crunching, devoid of "gut-feeling" ideological and cognitive biases—was going to evaluate defendants of all races and socioeconomic backgrounds on equal footing, based on data. In 2020, PJI reversed its position and petitioned for risk-assessment tools to be removed from pretrial justice systems. Their reason for the reversal? Instead of balancing the decision-making process, the algorithms perpetuated racial inequities. What happened? It turns out that the hypothesis was wrong. By assuming that an algorithm would eliminate racial disparities because it's just "looking at the data," we automatically surrender to a very important assumption: that the data is good. But what if, as it turned out, the data is already biased?

In 2014, in Coral Springs, Florida, near Fort Lauderdale, Brisha Borden and Sade Jones saw a Huffy bicycle and Razor scooter sitting unlocked as they walked to pick up Borden's god sister from school. They picked up the bicycle and scooter and proceeded to ride them. A woman who witnessed this alerted the police that someone had stolen the bike. Borden and Jones were arrested and charged with burglary and petty theft. The stolen items were valued at $80. Borden had a record for misdemeanors committed as a minor. In 2013, Vernon Prater shoplifted $86.35 worth of tools from Home Depot. He was a criminal with a long record, including an attempted-armed-robbery conviction and having served a five-year sentence for an armed robbery. What is interesting is that an algorithm predicted Borden, who is Black, to be at a high risk of committing future crimes, but Prater, who is white, was determined to be a low risk of committing a future crime.

It is difficult to prove why the algorithm was more lenient toward Prater than Borden. First, these algorithms are proprietary, so we can't really investigate how they are trained or how they function. Second, if the algorithm is a neural network, its decisions are buried in the neural connections, and for any given sample, we cannot trace an explanation from the input to the output. We know that the neural network produces an output for each input, but the output is driven by the weights of the connections, which have been adjusted during training, and it's impossible to define exactly what led to a specific prediction. Our inability to explain the output of a neural network is itself a major problem with the use of neural networks in the judicial system, but more on this later.

One possible explanation for this outcome is that the data set used to train the model was biased in a way that caused it to think Borden was more dangerous than Prater. To be clear, it's impossible to say exactly what happened in this one case (maybe it wasn't bias, maybe it was a glitch), but bias is certainly a possibility, and we should seriously consider the implications of using AI algorithms in criminal courts. Let's see one more example, and then we can discuss in depth the dangers of deploying these algorithms in the justice system.

This time, we look at the case of an algorithm used to alert police of possible future crimes. In 2013, Robert McDaniel was living in Chicago, in the Austin neighborhood, with his grandmother and adult siblings, when police officers showed up at his door. Unfortunately, Austin has one of the worst by-neighborhood murder rates in the city and some of the highest concentrations of gun-related crimes. At the time, McDaniel did not have a violent record. He had been arrested on a few occasions for marijuana- and gambling-related offenses.

This time, the officers were not there to arrest McDaniel; instead, they were there to tell him something that seems straight out of a dystopian sci-fi novel (possibly cowritten by George Orwell and Arthur C. Clarke). It turned out that an algorithm, run by the Chicago Police Department, had predicted that McDaniel would be involved in a shooting. The algorithm's prediction was informed by McDaniel's proximity to known shooters and shooting victims. The police officers told him that the algorithm didn't know whether McDaniel would be a shooter or a victim; it just knew that McDaniel would be involved in a shooting. And they were there to warn him that the Chicago Police Department would be watching his every move.

Take a moment to consider the situation. Imagine a police officer showing up at your door and telling you that an algorithm predicted that you will be involved in a crime, and therefore you will be placed under surveillance.

In 2017, McDaniel was shot outside his neighbor's house. Then, in August 2020, he was shot again in an alleyway near his house. Thankfully, he survived both attempts on his life. Well, the algorithm seems to have gotten it right. Indeed, he was involved in gun violence. But here is the problem: McDaniel had never had any serious trouble before the police showed up at his door in 2013. So is the algorithm really clairvoyant, or is there something else going on?

McDaniel has complained that the attention he started getting from police made people in his neighborhood suspicious that he was collaborating with them and labeled him a snitch. Again, it's difficult to say what exactly happened, and this book is not about investigating what in fact led to those shootings. But you can undoubtedly see how it's possible that added police attention in certain neighborhoods might contribute to some residents drawing the wrong conclusions about an individual. We're looking at this example to examine why algorithms might predict certain outputs and to evaluate the implications of those predictions. In the end, the algorithm was right: McDaniel was involved in a shooting (twice). But the prediction appears largely self-fulfilled. The irony is that the added surveillance that caused McDaniel unwanted attention and put his life in danger did not save him when he was being shot.

The problem with artificial intelligence in the cases we have discussed so far is that when an algorithm is used to predict the likelihood that you will commit a crime, the algorithm is not really analyzing you.

The algorithm was trained by analyzing a data set. The data set might contain a number of people that share a set of descriptive features that are similar to yours. For example, the samples in the data set might consist of people in the same neighborhood as you or in neighborhoods with similar socioeconomic profiles. The people in the training data set may be surrounded by gun violence, and many of them may have ended up committing gun violence. When the algorithm tries to fit your features to its knowledge base, your features may align well with others who have committed a crime or been involved in gun violence. But what if you are different? What if, while living amid poverty, desperation, and crime, you have not responded to those pressures in violent, unlawful ways? Then you better hope that the training data set involved an equal number of people in the same exact situation as you who have also managed to not commit a crime while living in a high-crime area. Unfortunately, we don't get to see how these algorithms are trained.

Whether we are trying to decide the risk of someone being involved in a crime based on their living situation and proximity to criminal actors, or trying to determine the likelihood that they will commit a crime in the future (as in the case of deciding bail qualifications), if we base our decisions on statistical models alone, then we are making our decisions based on the baggage that person brings simply by being placed in a category with people who have been involved in crime. There is something final—and gross—about this. Humans are not deterministic machines.[18] But the AI algorithms we have discussed thus far are. A human faced with a given problem may respond differently each time. But an AI algorithm that is trained and deployed will always provide the same response to the same input. This disparity between

---

18. There is a debate in academia about whether humans are in fact deterministic. Most scientists believe humans are deterministic and free will is simply an illusion produced by the complexity of predicting our behavior. Our behavior depends on the state of our system (mind, body) and the environment (the universe) in which we operate. Since we can't measure and interpret the state of every molecule in our universe to predict our behavior, then we appear to possess free will, and thus we appear nondeterministic.

humans and AI should at least be further explored before we entrust an algorithm with our future. The whole premise of the modern judicial process is that we would rather let a guilty person walk free than have an innocent person sent to prison.

Think about being in front of a judge, and let's look at two hypothetical scenarios. First, let's say that you did not commit a crime but are being accused of committing a crime; then let's consider the case where you committed a crime, you are remorseful, and you promise yourself that you will do your best to turn your life around and never do it again. Let's further grant the fact that a large percentage of criminals do indeed feel or appear to feel remorse during trial and make many promises to turn their lives around but in the end continue a life of crime. We must also grant the fact that there are many people who are surrounded by crime and are never involved in crime, and there is a small percentage of people (it doesn't matter how small) who turn their lives around after committing a crime. When you are in front of a judge, you hope that they will see past your baggage. You hope that you can convince them that regardless of what has happened in the past, you can still change your life; whether you have committed a crime or you are surrounded by criminals, you want to convince the judge (or the jury) that you are going to improve your situation and change your life. Now, what if instead of standing before a judge, you are facing a statistical model. Well then, there really is no more "you." There is *y'all* and what y'all did. There is no discussion of or pondering the future. There is only the past and what others in similar situations have done. The idea that people can change and that you can decide in a moment to change your life—that idea is gone.

The natural counterargument to the above discussion is that I am being naive, so let's address that. Indeed, judges can be biased, racist, or just plain incompetent. Just because you find yourself in front of a human judge does not mean you will get a fair judgment. This was the premise of the Pretrial Justice Institute's advocating for an algorithm to make bail judgments, to be impartial. But not all judges are the same, and not everyone is racist. Everyone has cognitive biases; however, some judges can see beyond circumstances and consider the impact of a sentence against the future of a person's life, and even when the evidence suggests a dark future, they can offer hope. But all statistical models, at least of the classes we are discussing, are similar in that they are not looking beyond your circumstances at a potential "you"; they are looking at data and making decisions based on the data they saw during training. This sounds straightforward, but we must again remember that the data itself is often biased. With a statistical model, there is no redemption; there is just history. If we lose the hope for redemption in the judicial process, however flawed it may already be, it will be a step in the wrong direction.

Before we move on from the criminal justice and policing use case, it is important to restate that these algorithms do not have a personal agenda; they do not intend to discriminate; all they do is analyze data. A particular problem with data is that many engineers do not spend enough time understanding the data they use because they assume that's the algorithm's job. But once we understand that the algorithms are looking at the data samples' features and reshaping, transforming, and projecting those features in different ways to find a predictive direction, we realize that the features that make up the samples are extremely important. By selecting a set of features around which to build a data set—for example, age, sex, race, living conditions (neighborhood, size of dwelling, number of adults per dwelling), income—we are already introducing biases into the system.

We are saying, "We don't know which of these features are the predictive ones—that's why we need a neural network. But we at least know that the answer must come from one of the features we have identified." This means that from the start we build a box to bound the problem of predicting an outcome. And that box is built around the initial features we select for our data set. We need to be very careful that the features we select are not pointing us in a discriminatory direction from the start. Now let's assume that we can build the perfect, balanced data set. Should we then automate aspects of the judicial process? Well, what do we mean by having a balanced data set? Balanced in which directions? Maybe we can finally agree that we have produced a data set that is not racially or socially biased in any way, but what about in terms of outlook? Does it have a means to identify individuals who, against all odds, can end the negative cycle and change their lives? Or is the data set myopic in its design such that it will simply identify the likelihood of committing a crime based on what a statistical majority of other samples in the same situations did? I think that if we can create an algorithm that demonstrates fairness over an entire population and that preserves the potential for individuals to hope and to redeem themselves in an instant—to the extent that is possible—then we will finally be able to automate judicial systems, because then we might have a mechanism that is better at making these discernments than humans. But we are certainly not there yet.

When it comes to employing neural networks in the courts, the biggest hurdle might be the *explainability problem*. As we have discussed, when neural networks make a prediction, we cannot trace the steps of the prediction back and explain the network's decision. It is not possible to argue and attempt to change the network's "mind." An integral part of our judicial process is that the accused can have representation and argue their version of the story—a much more difficult proposition if we cannot understand how the "judge" (i.e., the statistical model that has, in theory, replaced the human judge) arrived at its decision. You might be thinking, "What if the automation process that the judicial system used were to consist of multiple neural network models each arriving at an independent decision by analyzing the data?" It has long been known that a collection of models (in AI literature this is referred to as an *ensemble of models*) performs better than a single model. Indeed, it's likely that if the predictions of an ensemble of models agree, the prediction is better than if it was made by a single model. But note that "better" here means "better with respect to the data." An ensemble of models does not fix the data bias problem, and it still does not address the explainability problem. Even if several models agree that you are guilty, does that mean you are? Would you not want to fully understand exactly what criteria these models used in determining to convict you? Consider that multiple human judges can agree on your guilt, and you are still allowed to appeal their findings. You have several chances to do this, all the way to the Supreme Court.

WHY ACCURACY IS A FLAWED MEASURE

Finally, let's address the *accuracy problem*. It is not uncommon to read an article in the media that discusses the idea of accuracy in a machine-learning algorithm. Indeed, when it comes to algorithms aimed at replacing human judges in some parts of the judicial process, often the proponents of such algorithms boast the accuracy of the algorithms as greater than human judges in studies involving a large number of cases. The implied premise is that if an algorithm is capable of greater accuracy than a human, it would constitute an improvement; therefore, we should replace the human with the algorithm. On the surface, this sounds correct, but accuracy has a double meaning. The common definition of accuracy is tautological in that a high-accuracy measure is interpreted as a measure of correctness. If process A has higher accuracy than process B, then we assume process A to be more correct than process B and, crucially, to make fewer mistakes than process B.

In science, accuracy is a well-defined concept. It is the proportion of correct observations over all the observations. Suppose we have twenty criminal cases, and of those twenty cases, ten were correctly assessed by an artificial intelligence algorithm. The accuracy would be 10/20, or 50 percent. If nineteen cases were correctly assessed, the accuracy would be 19/20, or 95 percent. Studies have cited human judges' ability to predict recidivism—that is, the likelihood of a suspect to reoffend—as 60 percent accurate.[19] If we can come up with an algorithm that's 75 percent accurate, then it will already be performing better than humans, correct? It turns out that the answer is *not necessarily*. A measurement of accuracy does not tell us anything about the degree to which the algorithm generates false positives or false negatives.

For an algorithm predicting recidivism in suspects awaiting trial, a false positive prediction means that the algorithm identified a suspect as highly likely to reoffend, but the suspect never reoffended. A false negative prediction would be if the algorithm predicted that a suspect is very unlikely to reoffend, but the suspect went on to reoffend. Tables 4.1 and 4.2 give us an idea of the impact that false negative and false positive predictions can have on the accuracy of the predictions.

Table 4.1 Predicting and Observing Recidivism in One Hundred Cases (First Hypothetical)

Predicted behavior	Observed behavior	
	Suspect reoffended	Suspect did not reoffend
Suspect reoffends	35	5
Suspect does not reoffend	20	40

19. Edward Lempinen, "Algorithms Are Better than People in Predicting Recidivism, Study Says," *Berkeley News*, Feb. 14, 2020, https://news.berkeley.edu/2020/02/14/algorithms-are-better-than-people-in-predicting-recidivism-study-says/. For more details, please see the academic sources linked in the article cited, which are also included in the chapter sources at the end of the book.

Table 4.2 Predicting and Observing Recidivism in One Hundred Cases (Second Hypothetical)

	Observed behavior	
Predicted behavior	**Suspect reoffended**	**Suspect did not reoffend**
Suspect reoffends	35	20
Suspect does not reoffend	5	40

Table 4.1 describes a hypothetical study of recidivism in one hundred suspects awaiting trial. An algorithm analyzed those suspects and predicted that forty of them were likely to reoffend. During the pretrial period, thirty-five of those suspects went on to reoffend, and five did not. Similarly, the algorithm identified sixty suspects as unlikely to reoffend, but it got twenty predictions wrong; twenty of those suspects did in fact reoffend. Table 4.2 describes the same one hundred cases, but here the algorithm identified fifty-five suspects as likely to reoffend. Of those fifty-five suspects, twenty were incorrectly classified and did not end up reoffending. The algorithm also identified forty-five suspects as unlikely to reoffend, but five of those went on to reoffend.

The first thing we should note about these two different sets of results is that the accuracy is the same in both cases. Each time, the algorithm correctly classified seventy-five out of one hundred suspects, yielding 75 percent accuracy. The impact, however, that the algorithm has on the suspects and the overall population is quite different when comparing the two cases. In the first case (table 4.1), because of the algorithm's mistakes, twenty suspects were set free and were allowed to reoffend. In the second case (table 4.2), twenty suspects would have been kept in jail unnecessarily.

Clearly, an algorithm that makes no mistakes is preferable, but it is highly unlikely that we will ever have such a thing; even humans aren't great at predicting behavior. The point I am trying to make here is that accuracy is often overused when trying to indicate the effectiveness of an algorithm, but accuracy alone doesn't always tell the whole story. In use cases or situations where the results of an algorithm carry a large weight of responsibility (e.g., in the judicial process or in health care), it is very important that, in addition to accuracy, we discuss the types of mistakes the algorithm makes. In the hypothetical recidivism example just discussed, it's difficult to say a priori which is a better result. If the choices are between an algorithm that's likely to err on the side of releasing suspects who might reoffend or a stricter algorithm that's likely to keep more suspects in jail than necessary, the right choice might depend on the types of crimes the suspects are accused of having committed. Consideration should also be given to, for instance, how an individual being kept in jail affects their family versus the risk of reoffense. If nothing else, this example illustrates the nuances involved in the process of predicting human behavior and the cost of mistakes in that process. When debating the validity of replacing human judges with automation, we can't be seduced simply by the prospect of higher accuracy.

Before we move on from the discussion of AI in the judicial process, it's important to make something clear. While it's understood that human judges are not perfect (they can be prejudiced, can make mistakes, and indeed can be biased), the concern with a premature migration to automation is threefold. First, to replace an existing system with a new one, the new system should be clearly better than the old one. Second, because automation algorithms fundamentally work by analyzing data, many people feel inclined to trust their results and find it much more difficult to disagree with or dispute the results of a machine-learning algorithm than those of a human. Studies have documented human decision-makers' inability to oppose recommendations made by automated systems. This observed behavior is in part born out of an already biased and ingrained acceptance of a machine's ability to outperform us in every setting, from mathematical calculations to planning and intelligence. (This clearly feeds the fears of AI's world domination. AI systems are not just out to get us; they are also more intelligent and more capable than we are.) But, in addition, we have developed an element of complacency over time as we have witnessed strings of automated successes and gained confidence in machine performance. Still, while automation systems may succeed most of the time, the edge cases, when they fail, may be devastating. Therefore, we need to be fully confident that the algorithms are indeed an improvement over a human in every possible way. Lastly, until our algorithms can intuit potential and predict change in an individual regardless of their past, the rest won't matter. It will not matter how accurate they are, and it will not matter how balanced their false positive and false negative mistakes are. They will nevertheless be nudging us toward a less hopeful future, without redemption. Clearly, there is a lot that must change in the way we interact with and discuss AI before we can entrust it with our freedom.

ADVERTISING WITH ARTIFICIAL INTELLIGENCE

We have been examining current applications of AI that require more nuanced conversation and awareness from the public. These are areas of immediate and real concern in the application of current AI techniques. Another application of AI that should be generating a lot more discourse is advertising (or any area where AI systems can influence our behavior). In the previous section, we discussed bias as the main concern with AI algorithms. In this section, we center our concerns on the method by which these algorithms learn: optimization.

Recall from chapter 3 that machine-learning algorithms learn by optimizing some objective function. The objective function is a method by which the learning process can measure its progress toward a specific goal. In some problem domains, the goal is to minimize a loss function—for example, to minimize the difference between the model's output and the ground truth. In other cases, such as the reinforcement learning briefly discussed in this chapter, the goal might be to maximize a certain reward function. Invariably, however, the system can be framed as an optimization problem where the goal is always to reach an optimal value.

Let's use an example from chapter 1: handwritten-digit recognition. Suppose we want to train a model to recognize handwritten digits and classify them as 0, 1, 2, . . ., 9. If you recall, we can use a training data set of 60,000 images of handwritten digits that are all labeled with the ground truth: images of digit 0 contain a label of 0, images of digit 1 contain a label of 1, and so on. During the initial stages of training, the model won't be very good, so it might predict that an image of a 5 is a 2, for example. The training process then measures a loss (a sophisticated difference measurement) between the predicted value and the ground truth and adjusts the model to improve its predictions in subsequent runs. Every adjustment that the training process makes to the model is done to optimize a certain objective: to correctly recognize the digits. Optimization problems have been shown to be extremely powerful at slowly shaping a model to be more and more effective at achieving some goal. What if, instead of predicting the correct label for an image of a digit, the goal of the model is to influence us in a certain way? For example, what if the goal of an algorithm is to get us to buy more products of a certain brand or from a particular online store?

We like to think of ourselves as obdurate, strong-minded, and rational individuals who deliberate on the reasons for our behavior. We like to think that we have our own opinions and needs, and we are, to quote a great sitcom, "the masters of our own domain." But the truth is that we are incredibly malleable and, en masse, not all that difficult to influence.

Suppose the goal of an algorithm is to get you to buy a shirt from a certain store. The algorithm may send you a few shirt ads. You might see the same ad once a day for a couple of days and not click on it. Maybe on the fifth day, you click on the ad and decide to check out that darn shirt and see what's so special about it. The algorithm might learn that it takes you a while to click on an ad, but eventually you do, so in the future, before relenting or changing tactics, it may stick to the same strategy for at least five days.

Now let's say you visit the site, but you still don't buy anything. How does the algorithm know about you anyway? Why is it targeting you? Remember that free coupon you got a few months ago for cotton socks? Your friend gave you the coupon as part of the store's "add a friend" campaign. That's how the algorithm knows you, and that's how it knows of your connection to your friend. Don't worry, the algorithm is also sending ads to your friend, and it turns out that your friend loves that shirt and buys it right away. Now the algorithm tells you that your friend just bought an awesome shirt and asks, "Wouldn't you like to keep up with your friend and get a shirt too?" If you buy it this time, the algorithm incorporates that information and now knows a way that worked at least once to get you to buy something.

This is just an example, and of course not everyone cares about what their friend buys, but some people do. And more importantly, each time the algorithm interacts with you, it is measuring your response, and it is adjusting itself to maximize the likelihood that in your next interaction, you will react the way it wants. The truth is that, for as intelligent and self-aware as we think we are, we don't stand a chance against the relentless power of an optimization algorithm. Now think of the most vulnerable section of the population when it comes to influencing. Is it truly any wonder that anxiety and mental health issues in teens are at an all-time high?

There is no question that advertising works. Companies spend billions of dollars every year on marketing campaigns because they understand that if they find the right connection between a large part of the population and their product, they stand a higher chance of selling that product. Most of the time, this connection is an emotional connection. Think about some of the most successful ad campaigns from companies like Apple, Coca-Cola, or Nike. They spend very little time, if any, speaking about the product and what it does. Instead, their ads are about lifestyle. They show their product being used by an elite athlete or a celebrity in some idyllic location. They make you feel like maybe if you had that product, you could also have an awesome life and spend your winters skiing in the Swiss Alps.

Now, of course, not everyone is influenced by the same ad. And some people are harder to influence than others. Classically, ad campaigns were informed by focus groups. The focus groups were small groups of people that were meant to represent a certain sample of the population. Based on how the focus group reacted to the ad campaign, the companies would extrapolate on how successful the campaign would be on the larger population. But focus groups aren't perfect representations of a large population, and so the data they gathered from these groups was noisy. But what if they had a method to tailor ads or entire campaigns to smaller directed groups of people or even specific individuals? With the algorithms we have discussed so far, it is possible to send ads to specific individuals based on their online shopping history. (In fact, it happens every day.) It is then possible for the algorithm to learn strategies that work to get you, specifically, to maximize your spending habits.

You may think, "Well sure, sometimes I listen to these suggestive ads, but if I buy the product, it's because I needed it anyway; it's not because some algorithm told me to." Let's test that hypothesis. Think of all the products you have bought in the last year that were suggested by your favorite online store. How often did you end up using it, and did it really improve your life? Again, it's quite possible that for you specifically we are not there yet, and maybe you are impossible to influence, but most of the population is not like that. In fact, companies are now spending billions of dollars on researching and developing these types of recommender algorithms, because they see the impact such algorithms are having on spending habits.

Perhaps shopping isn't your thing. Maybe you use streaming services to watch your favorite movies or TV shows. Have you noticed that when you log in to your favorite streaming service, your landing page is different from your grandma's? Last week, you decided to watch a couple of true-crime documentaries, and now when you search documentaries, they are mostly about true-crime stories. How often have you sat down with the idea to watch only one or two episodes of your favorite series only to spend your entire Saturday binging the whole series? The types of algorithms that power these services and suggest what new show or movie you should watch based on your history are uninspiringly called *recommender systems*. Their job is to maximize your time watching something. We also see these algorithms on news sites. In recent years, we have seen the impact recommender systems can have on dividing a population by creating so-called echo chambers. This describes what happens when recommender systems funnel like-minded people into the same corners of the internet, away and protected from differing opinions or facts.

If you start searching for flat-earth ideas, you will progressively get more stories or comments about flat-earth conspiracies. But what's interesting is that you will also get more suggestions to read about other conspiracies, like Big Foot or Pizzagate. Eventually, some people fall down the rabbit hole of conspiracy theories, and because the same algorithms have found other people with similar proclivities for conspiracies, they all start building on each other's comments, and there is no one in the group with a different opinion to offer some perspective. The algorithm is also not going to suggest material disputing the flat-earth idea because the goal of the algorithm is not to teach or to differentiate between facts and fiction. The goal of the algorithm is to maximize traffic to the "news" site. And maximizing something is one of the things that algorithms can do very well.

When it comes to algorithms that interact with humans with the goal of then influencing human decisions, we need to be very careful. In these systems, the human becomes another variable, and the root of the problem is that the goal of the algorithm is not aligned to the well-being of the human. The amount of disinformation and division in the world today is in no small part due to how easily we can be influenced by recommender systems. Propaganda and division aren't new forces we have to contend with. They have always existed. The problem we have today is scale. As our algorithms become more powerful and more capable of tuning into our fears and insecurities to drive our behavior, the more profound their influence will be, which can have catastrophic consequences—such as perpetuating biases and inequalities or maximizing a company's profits at the expense of our mental health or compromising our ability to discern fact from fiction. This—not robot overlords—is what worries me about AI.

○ ● ○

Why have we spent so much time discussing biases and critiquing AI in this chapter? Shouldn't a book about AI promote its benefits?

The purpose of this book has been, first and foremost, to inform. Its aim: to explain how some common AI algorithms work and to shift the conversation from hypothetical future problems that we don't know how to solve—problems that, indeed, we don't know when, how, or if they will manifest—to current problems we *must* be discussing. Now that we have covered some chief areas of concern, we can ask whether it is all doom and gloom or whether there is hope. Are there positive reasons to pursue AI?

I spend a lot of time researching artificial intelligence algorithms, and the primary reason I wrote this book is because I wanted to share the fascinating history of artificial intelligence and the elegant mathematics behind its development. These algorithms and their inner workings are too remarkable to be known about or understood only by a small group of academics and engineers. To be fair, this exclusivity isn't unique to artificial intelligence. Much of the beauty and elegance in science only a few get to see. We don't build galleries dedicated to theorems and proofs. It's easy to blame the public for their lack of interest in the sciences simply because "it's hard." But a fair bit of blame should also go to the scientists who obfuscate and convolute information with arcane phrases and difficult-to-understand language.

We can appreciate science at many varied levels. We can become scientists and understand every technical aspect of a theory, or we can forgo delving into the technicalities and learn about a theory's implications and how it might apply to our lives. In a similar way, when we go to an art museum and gaze at a Renoir painting, we don't need to be an artist or even know how to hold a paintbrush to enjoy the work. Sure, an artist might be able to appreciate the different brushstrokes or understand that the paint itself and how it has been prepared are integral parts of a masterpiece. Great artists mix their own paints by suspending different amounts of pigment in oil. This creates varying degrees of translucency in a painting. An artist can view a masterpiece and recognize these details in the work, just as they grasp perspective and how 3D objects can be projected onto a canvas without losing the spatial relationships in a scene. But those of us who don't understand these details can still derive enjoyment from viewing art. We can still marvel at a beautiful painting and what someone can do with a bit of color and a brush. This is an important part of being human—appreciating and taking pride in what we can accomplish as a species. Even if we personally can't sculpt the great David, we can at least enjoy knowing that someone could. Much of the beauty and elegance in the sciences is accessible only to those who elect to study those subjects, and most people live entire lives without being exposed to those qualities, without having the chance to marvel at what we as a species have achieved. That is unfortunate because we only have one life to live. I believe there is a lot of beauty and elegance in artificial intelligence, and the principles behind these systems are simple enough that they can be understood by anyone. Our discussion of fears and the downsides of AI can be misconstrued as a certain apprehension on my part or distrust of AI. So I think it bears clarifying where I stand on AI and the future.

I consider AI one of the most interesting areas of computer science today. Due to relatively recent advancements and its capacity to affect our lives, it gets more attention from the media and the public than any other subject in computer science. When concerns about AI are brought up, they almost inevitably involve big existential threats. They involve AI systems gaining awareness and somehow formulating a plan to get rid of us, an inferior species. It's impossible to say whether AI algorithms will ever become self-aware and generate their own agendas and "free will." It's impossible mostly because we do not yet have a scientific theory of consciousness. We do not understand how it arose in us or how it could arise in others, so today it seems implausible to create an artificial system truly capable of being conscious when we don't really know how it could be done. Let's pause and allow that thought to sink in. As we have seen, our AI algorithms today are more "A" than "I." We can stop worrying about a robotic revolution for just a minute. Despite all the hype, we don't yet know enough to create a conscious AI; this danger is not imminent.

Of course, we also don't know how long it will be before a theory arises that changes everything and we get self-aware robots. The best we can do today is start having conversations about the ethics of conscious AI. If it ever becomes a reality, how should we treat it? What responsibilities should we bestow on it, and what are the risks of making it responsible for running vital systems? If such systems are truly conscious, is it cruel to have them endlessly perform mundane tasks? And are we allowed to shut them down when we no longer need them, or is that like killing an intelligent being? What are the rights of artificial conscious systems? We should have these conversations before anyone attempts to create consciousness, and failure to have a plan in place could create whole new areas of unexpected problems and threats.

Lastly, we must address a fundamental concern of AI—which might be exacerbated by a conscious AI in the future. This concern is known as the *alignment problem*. The alignment problem describes the phenomenon where the path to reaching the goal for an AI system, even if the goal is initially defined by humans, may not be aligned with our own well-being. Stuart Russell, a British computer scientist, in his 2019 book *Human Compatible: Artificial Intelligence and the Problem of Control*, postulates a set of possible scenarios where an AI system could very well destroy humanity in the process of trying to accomplish goals set out by humans. In one such example, he imagines a geoengineering robot that is tasked with deacidifying the oceans. In the process of deacidifying the oceans, the robot devises a plan to use up all the oxygen in the atmosphere: an unfortunate side effect being the death of all oxygen-dependent organisms on earth, including us.[20] This example illustrates the problem of explicitly defining goals for an AI system that shares very little with us. Yes, we might design these systems, but they operate in ways that are very different from our own way of functioning. Although our algorithms come from our own imaginations, future conscious AI systems are unlikely to share our values, concerns, and ultimate goals. This last part is subtle. Even in Russell's example, superficially it appears that humans and AI share the same goal, deacidifying the oceans. But at a fundamental level, our goal is to survive and prosper.

We can imagine the alignment problem as existing on a sliding scale where we start at one end and the issue of alignment exists in an immediate time frame. We discussed these issues in the first half of

---

20. Stuart Russell's examples of bad manifestations of the alignment problem are not necessarily specific to conscious AI. Indeed, all the problems we have been discussing thus far—AI in the judicial process, advertising, and so on—constitute weaker forms of the alignment problem. But with conscious AI systems, the problems would be greater since presumably these would be more advanced systems, more integrated into our society, with more freedoms to move and make changes to and decisions about our lives and future.

this chapter, where the algorithms we are using are trying to maximize the likelihood of some desired outcome based on an optimization process. Due to the simplicity of the algorithms and the complexity of the environments in which we sometimes deploy them, however, a *successful* outcome as discovered by the optimization process can ultimately be harmful to us.

The further we move to the right, say, on the sliding scale, the more dire the consequences of misalignment between humanity's and the AI's goals. Eliezer Yudkowsky has been warning the world about what he perceives as the negative implications of artificial general intelligence (AGI) for years now. AGIs are algorithms with the capability to adapt and learn—autonomously—to solve any class of problems that human beings can solve. In Yudkowsky's views, this process risks running out of control and leading to "artificial superintelligence." Surely, AGI algorithms are still far beyond our current capabilities; nonetheless, the thought of an AI seeing us in the same light that we see chimps, as interesting creatures but far too primitive to hold an *intelligent* conversation with, must be unsettling. In a blog post where he explains his "death with dignity" strategy, Yudkowsky goes into detail concerning the dangers we face. But whereas he seems to suggest that it may already be too late to stop the eventual machine takeover, I hold a more positive view. After all, we are still in charge. We still (as humanity) get to decide where we deploy these systems. Any problems we run into today with AI are purely self-inflicted. We should, however, understand under which conditions we might lose control and reach the point where it will be too late.

The alignment problem is fundamentally the same whether we are discussing an AGI or the AI algorithms we have visited thus far; the only difference is the scale of the problem. The more powerful the algorithms, the larger the landscape of possible responses leading to situations that are not beneficial to us. It is important to note that AGIs are not necessarily sentient or conscious systems. It is entirely conceivable to reach an artificial superintelligence without generating consciousness. In this sense, the danger is not in the algorithm being out to get us. The danger is simply in us being in the way.

If we continue to move our imaginary slider to the right, we can imagine self-aware and conscious systems at the far end of the sliding scale. If we take a moment to carefully consider the conscious human mind in contrast with an AI algorithm (conscious or not) trained to maximize a specific goal, we will notice an important distinction that makes all the difference. In our thought experiment, we can imagine that our conscious process might be driven by a set of objective functions similar in principle to those driving our AI algorithms. But unlike the AI algorithms we designed, the objective functions driving our conscious process have been honed by millions of years of evolution. We do not have a single objective function that we must maximize; we may have hierarchies of functions ensuring that our pursuit of higher-level goals does not ultimately end in our own demise. To take an example from Nick Bostrom's 2012 article "The Superintelligent Will: Motivation and Instrumental Rationality in Advanced Artificial Agents," an AI algorithm purposed with maximizing the number of paper clips in the universe may find that humans stand directly in the way of achieving that goal. It is us or paper clips, and the AI algorithm is not compelled to choose us. A human provided with the same goal of maximizing the number of paper clips in the world might arrive at the same conclusion but avoid wiping out humanity because of competing goals from lower-level objective functions linked to our own survival.

Consider AI reaching a level of sophistication where its objective functions are complex hierarchies driven by some evolutionary process (or design) that fundamentally aims to maximize the survival of the algorithm. The more power we grant to these algorithms—connecting them to our power grid, essential services, internet, military—the more difficult it will be to control them. The complexity of their objective functions may be such that the algorithms are indistinguishable from conscious beings, but crucially, their goals and value systems, not having gone through our evolutionary processes and experiences (which drove us to control our violence, collaborate, and trust each other), will not resemble ours. This is the point where we might finally lose control of a much more intelligent algorithm that, now, also cares about its own survival.

The alignment problem encapsulates all such cases where, even after we have carefully described a set of goals, the fundamental differences between us and AI systems mean that even when the AI systems are not purposefully out to get us, our realities are so distinct that intrinsic, undefined outcomes will not align to a successful cohabitation. The funny thing about the alignment problem—often reserved for use as an ominous warning against AI rebellion—is that it is hardly special. It is simply a question of where along the sliding scale we find ourselves. Today we are still near the very beginning. We do not have AGIs, and we will not have sentient machines anytime soon, at least not until we agree on what "sentience" or "consciousness" even means. But we must be aware of the alignment problem at each time step. That is the message of this book. We can worry about the problems of the future, but we certainly need to understand the problems we face today!

And while there is a lot to consider and to worry about with respect to AI, and important risks that must be addressed, consciousness and AGIs are not the only threats. That was the point of the first half of this chapter—to encourage an honest conversation about the systems we have today and understand that we are already using our very primitive AI systems in ways that can harm us and are harming us. Instead of only focusing on a future plausible threat, perhaps we should also focus on the current ones, even if they are less exciting.

Once we are aware of the problems and have a basic understanding of how these systems work, we can start to differentiate between reality and fiction. We can see that, like any tool in our history, AI can be helpful, and it can be damaging. Having these discussions should empower us to believe that we can select the beneficial use cases and drop the harmful ones, instead of fearing all aspects of AI and deciding that we want to ban it (as if that were possible). That's why researchers need to be honest and the public needs to be informed. History has shown that once we discover a tool that can be helpful, we are not capable of abandoning it even if it poses significant risks. My hope and my goal have been to help expose the benefits and the risks of AI, especially of our current AI systems, to attempt to combat irrational fears and replace them with the need to formulate helpful actions.

CONCLUSION:
FROM CHAT ROOMS
TO CHATBOTS

W e have talked about the drawbacks, so let's close by discussing the advantages to adopting AI solutions to various problems we face. Where can AI help? Are there immediate benefits to AI, or is it just a dangerous novelty? As the technology matures and our ability to employ these technologies in more areas increases, we will find benefits we are not yet able to imagine. This is the nature of discovery. When the personal computer was invented, no one could have predicted all the different use cases and solutions they represent today. As an exercise, we will focus on two specific areas of AI research when discussing its benefits, but note that we are just scratching the surface of what AI can do, and its benefits will only increase with time. Our areas of focus are autonomy and health care. By autonomy, I mean systems that operate without human intervention and perform tasks on behalf of a human. Examples include robots operating in warehouse environments and self-driving cars. The health care sector is broad (and expanding), and the potential for AI interventions is high.

Self-driving cars are already operating to some extent on our highways. For the most part, these cars are not yet entirely autonomous, and they require human supervision and intervention to prevent accidents. There are companies like Waymo, which operates an urban fleet of entirely self-driven taxis. In some cities where they have been running for years, mapping different routes, the cars don't even include a safety driver (a human driver used as a last resort to avoid an accident). There are also companies that are employing smaller self-driven vehicles for food delivery. These operate in urban cores, along predefined routes and amid more controlled environments. When we think of self-driving cars, however, we usually think of our own cars taking us to different places, different cities, different countries, all without our having to touch a steering wheel. Companies like Tesla believe they are close to making this happen. Other people in the industry are more skeptical and believe the technology is still not good enough to achieve full autonomy for at least a decade. Whichever side you take, one thing is true: self-driving cars are coming; it's just a matter of when. We then must ask the question: Is this a good thing?

Let's be honest, when we drive down a busy highway on any given day, and we see how some people drive—either at a snail's pace, clogging the lanes, or too fast, weaving in and out of traffic—it's difficult to argue that, in principle, this *finesse* is hard to automate. I think that, in the future, when all cars are self-driven and the roads and infrastructure are in place for car-to-car and car-to-infrastructure communication, traveling by car will be much safer than it is today. We will no longer have to worry about impaired driving. When cars can communicate their moves to other cars in their vicinity, the flow will be much smoother, and traffic jams should be less common. Think of our reaction times versus the reaction times of machines. Most driving manuals suggest we should drive at least two seconds behind the car in front of us. The reason is that two seconds should give us enough time to react to a sudden move by the car in front. Computers can react much more quickly than that; therefore, self-driving cars should be able to close that gap safely so that, even on crowded highways, traffic should flow less impeded than it does today.

Think of what it means for a car to no longer require a human driver. Visually impaired people will have more freedom to move around and run errands without requiring assistance or hiring taxis.

Single parents who struggle with balancing work and picking their kids up from school will be able to have their cars take them to and from school—all without a driver inside. The question is whether all of this can be done safely. There is a whole industry grappling with this question right now. I think the answer is that for self-driving cars to become the standard mode of transportation, they don't have to be perfect; they simply have to be safer than human drivers are today, and I think that's certainly possible.

Clearly, none of this is easy. And there truly is a whole industry today discussing and defining the safety measures that will create the self-driven cars and necessary infrastructure of tomorrow. Some of these discussions involve insurance: Who is responsible for an accident when there is no human driver? Some are around ethics: How does a car decide between crashing into another car with people in it or veering into a median in a solo collision? These are all important questions, and they are well beyond the scope of our book. Our goal here is to understand, as it relates to AI and assuming that all ancillary questions are answered and resolved, why it is that self-driving cars should be automatable and a good use case of AI whereas the judicial process is not. To me, the test for whether AI is a good fit to solve a human-related problem is if it doesn't aim to influence our behavior (as in the case of advertising and recommender systems) and if there is a fundamental set of rules to solve the problem.

In the case of criminal courts, what happens to suspects, how they should be judged, how long they should be in prison, and the severity of the sentence are more nuanced considerations than a strict set of rules. These are decisions that require compassion, and they require a judge to sometimes err on the side of hope to end the cycle of violence and transgressions that got the suspect into court in the first place. Sure, this process leaves gaps for terrible people to take advantage of the system, but remember that we have already accepted this as the price for our freedom. This is the very principle of our court systems, that it is better for a guilty person to walk free than to send an innocent person to prison.

So why is self-driving different? The problem of self-driving is fundamentally a perception and reaction problem. If we look at the basics of driving, a set of overarching questions emerges: Is there an obstacle in front of me, and if so, how do I react to it? Do I stop, or do I avoid it? To make those decisions, the cars have an array of sensors and algorithms that interpret the input from those sensors and correct the car's position and speed accordingly. As long as the goal of the algorithms remains, ultimately, to avoid a collision (as is the primary goal for humans), I believe AI will eventually be capable of outperforming human drivers. The problem we need to avoid is asking too much from an AI system—for example, expecting it to not only avoid a collision but also, considering that a collision is about to happen, to choose the best outcome for a collision. This is where things could turn bad quickly.

It's hard to define what "best" means, and "best" for an algorithm might be different from "best" for the passengers inside the vehicle or for those outside. Consider the hypothetical case of a self-driving AI trained on a data set of collisions and outcomes with the goal of learning to choose the best outcome given that a collision is imminent. Most modern cars are generally safe and are comparably adept at keeping passengers safe during a collision, so let's assume that in this data set the most salient consequence of a collision is insurance cost. In an imminent collision when the AI's choice is between crashing into a luxurious car versus an older one, the AI might choose to crash into the older one to minimize the financial cost of the collision. Clearly, this leads down the wrong path of an ethics argument.

What should the car do instead if it knows that whatever it chooses, it's going to crash into something? I don't think I have the answers, and these unresolved questions are part of the reason we don't have fully autonomous cars yet. For the purposes of our discussion, let's say that an answer could be to reduce speed and crash into the object in front at a slower speed. Is this always a good solution? Probably not. But at least it's deterministic, and it avoids the problem of the AI choosing the cost of a collision by putting a price on people. Indeed, an AI that is capable of driving better than a human is a difficult problem to solve, but I think it's possible because, when we consider how we drive, fundamentally we are following a set of rules. We are primarily trying to avoid objects, and in most cases, when we get into a collision, we hit an obstacle while trying to avoid other obstacles; in general, we are not making ethical or financial choices for what's best to crash into. I suspect that it is possible to design an algorithm that, while not perfect, can learn enough rules to avoid more collisions than humans do and can opt to protect its own passengers when it gets into a collision, instead of weighing which other people it is best to crash into.

AI algorithms are currently being deployed in another important domain where their applicability will only increase in the future: health care. One of the biggest bottlenecks in many health care systems around the world is in diagnosing diseases. AI systems like the convolutional neural networks we discussed in chapter 2 can be used to analyze images of disease and grade the severity of the disease in patients. Much of the current work is still in the research stage, but many use cases already abound. For example, technicians can use neural networks to classify chest X-rays and distinguish between scans showing signs of pneumonia and scans showing healthy lungs. Also, engineers have trained neural networks to analyze images of potentially cancerous biopsied tissue, including classifying and detecting varying grades of melanoma (skin cancer), prostate cancer, bladder cancer, and many other types of cancers.

The networks are trained with data sets of images labeled by expert pathologists as exhibiting certain grades of cancer. During the training phase, the system learns to pick up on the features of the images that might indicate different severities of cancer. For now, most of this work is being conducted in research labs, but the hope is that eventually automation systems could be deployed to help ease the burden of diagnosis on medical experts. It's well known that the best chances of surviving cancer depend on our ability to find it early. This means that the quicker an expert can analyze a test from a patient, the sooner the patient can get the help they need. Unfortunately, experts aren't always available when they are needed, and the ones who are available are stretched thin and overburdened with a long line of waiting patients. We are not there yet, but these automated systems should help greatly augment the diagnostic power of hospitals and labs around the world.

A subfield within the health care domain that looks to benefit greatly from the application of artificial intelligence to its research is in the development of pharmaceuticals. Alphabet, the company that owns Google, recently spun off a brand-new company called Isomorphic Labs, dedicated to helping find treatments for diseases related to proteins and their shapes. It turns out that many terrible degenerative diseases—like Alzheimer's, Parkinson's, and Huntington's—are suspected to be related to misfolded proteins. Proteins are made up of long chains of amino acid molecules. Ribosomes, which are molecular machines found in all living cells, fold protein strands into specific 3D shapes. These shapes are responsible for much of the protein's functioning. Sometimes, the folding process breaks down, and the protein is folded into a shape that affects the expected function of the protein, resulting in what is known as *protein misfold*. Understanding the shape into which proteins are folded in 3D space is vital to designing drugs to treat the effects of misfolded proteins. DeepMind, an AI research group at Google, created an algorithm named AlphaFold that is trained to predict the shapes of proteins. In the relatively short time this algorithm has been training, it has reportedly outperformed the speed and capabilities of human research teams in predicting protein shapes. Isomorphic Labs was created to further expand on the research that produced AlphaFold and to eventually work with pharmaceutical companies to produce drugs that can treat such diseases.

These different health care scenarios are great examples of uses cases where in principle AI algorithms should benefit us. Why? Because diseases and their diagnoses typically follow a set of rules. For example, depending on certain characteristics of a group of cells in a biopsy slide, the shape of the cells, and the number of cells in a given region, as well as sudden unexpected blood vessels feeding these cell regions, a pathologist might determine that the tissue shows signs of cancer. The rules are complicated, and humans don't know all the rules in most cases; this is why the process of diagnosing a patient is complex and requires experts. And it's not 100 percent accurate—in fact, for some cancers, the accepted detection accuracy for human pathologists is around 75 percent.

This is also why we need neural networks to discover the rules for diagnosing diseases. Even an expert pathologist whose job is to detect these diseases will produce diagnoses that often differ from other experts' assessments. Because the rules for diagnosing a disease are complex and not well defined, pathologists learn them through experience and by developing what might be called an "instinct" for it. If the rules were well understood and well defined, we wouldn't need artificial neural networks to discover the rules. We could create a classic decision tree, which could check a list of features and, depending on which features are "ticked," could decide whether the disease is present. But this is not the case. Instead, we need to discover the rules through the learning process of artificial neural networks. Although the rules are complex and we have not yet discovered all of them to systematically diagnose every disease, it should be possible to discover them.

Determining whether a patient has a disease should not be, at its core, subjective and nuanced. It must be a matter of learning to detect the signs of the disease. This is why artificial intelligence algorithms are in principle a good fit. The process should not require the algorithm to make ethical or compassionate decisions. The process simply trains the algorithm to learn to identify features that predict a certain disease. Of course, researchers cannot be careless about the data or the process we use to train these algorithms. Researchers must indeed be careful that biases are not introduced in the training process, and they must be extremely careful to not draw broadly optimistic conclusions about preliminary results, which might lead to prematurely deploying systems that don't generalize well over an entire population. This could endanger the larger population by providing incorrect diagnoses and further erode the public's trust in these systems. But nevertheless, in principle, diagnosing a disease in a patient should be automatable because it is not a subjective problem.

We are just beginning to understand and cultivate the potential benefits for AI in health care. Automated disease diagnosis, discovery of new drugs, biomedical implants meant to monitor specific body functions—all are areas where AI research has the potential to change our lives for the better. Elon Musk has started a new company called Neuralink, which is researching methods for implantable brain-machine interfaces that, combined with prosthetics, could help disabled people gain lost functionality. Some researchers hope that brain implants will one day help cure specific types of blindness caused by damaged regions of the brain. These examples should give us hope and make us optimistic about the future of AI. AI can benefit us, and it can harm us; it's all about how we choose to use it.

Earlier, I said that self-driving cars and robots that work in warehouses stocking and dispatching items are examples of autonomous AI systems. I went ahead and explained self-driving cars but did not elaborate on robots in warehouses. Companies today are already employing robots tasked with filling skids with boxes, moving the skids to stock shelves, and retrieving skids from the shelves to different loading areas for dispatching. These robots exemplify AI systems that operate well autonomously, without human interaction. They can identify the objects they are tasked with handling. They physically move the objects from location to location, and they can avoid obstacles in their path.

Some see these robots as a positive example of machines performing what are among the most dangerous jobs in warehouses and removing humans from areas of frequent accidents. Robots don't need weekends off or vacation or sick leave, and they do not require sleep. So from the point of view of the company owner, they are better at scaling production. A clear benefit of this is an expanding economy. It is also important to note, however, that automation leads to displacing humans from jobs they need, which many fear will add to an eventually catastrophic mass unemployment. Whenever new transformational technologies arise, especially technologies capable of automating tasks traditionally done by humans, job losses will inevitably follow. I do not know whether the current path of AI and automation will lead to catastrophic mass unemployment. Some jobs will certainly disappear, but I suspect that new ones will be created to support new industries. When we look at the history of civilization, there have been many periods of invention and automation and job displacement, but new industries have sprung up to provide new job sectors as well. What will likely happen is that jobs will continue to become more and more sophisticated, requiring higher education. The sector of the population caught in the change will lose their jobs, but they should not be left by the wayside.

Such shifts, and especially their consequences, are no longer a problem of AI or technology; they are a matter of government policy. It would be quite unfortunate were we to start advocating for the constraint of technological advances simply because we can't take care of the population affected by the shifting job market. Some thought leaders in these areas suggest universal basic income as a possible solution to the job losses created by automation. I do not know whether this is a good solution; what I know is that a good and healthy society should aim to take care of those in need. How we provide help to people displaced by automation—whether it is by offering a type of universal basic income or retraining opportunities in emerging industries or a combination of those possible solutions—will be determined by government policy. When it comes to policy, we ultimately make the decisions.

○ ● ○

In the 1970s and 1980s, only a select few technologists were familiar with computers and their capabilities. Today, most people understand what a computer is and what it can do. Of course, an expert will still know more than the average person, but a conversation about a software application no longer must begin by explaining what a computer is. Similarly, we can discuss the internet as a great tool for staying connected with friends and family, as a tool for learning about the world and different cultures. The internet, however, also presents dangers.

When we discuss these qualities, the good and the bad of the internet, we can move straight to the problem without the need to begin with an explanation of what the technology is. We may suppose that when the internet was in its infancy in the early 1990s, a few tech visionaries could have predicted a future where the internet would be central to our lives—a future where we would work, hang out, seek refuge and entertainment, and indeed live on the internet. But discussions addressing the advantages and disadvantages of such a future would have been difficult at that time.

If you are old enough, you may recall that during the nineties chat rooms became a sensation. You could type a message, and a stranger halfway around the world would reply instantly! As a kid, I spent hours in these chat rooms. In those days, many people—including most of our parents—did not understand what a chat room was. Most people still didn't really understand what you could do with a single computer, let alone what you could do with a network of interconnected computers all over the world. It took nearly three decades to get to any meaningful conversation about the internet—discussions that we have only recently begun concerning child safety, privacy, online identity, security, and so on. We have had to wait for internet (and computer) literacy to reach a point where we no longer must explain what the technology is so that we can discuss how it is affecting us.

When it comes to artificial intelligence, the conversation has not yet matured to this level. We may have digital assistants all over our houses, replying to our endless queries, but we don't yet understand how such systems work, what information they are collecting, or how they might use that information in ways that can affect our lives. Our kids are having conversations with these digital assistants—getting them to tell jokes, asking questions about animals, and requesting videos on YouTube. We may find this as amusing as the chat rooms of the nineties. But this time, we shouldn't wait three decades to discuss the responsible use of this new type of technology.

The hope is that—with this book and others like it—the public will gain enough understanding of what artificial intelligence is and its basic capabilities to appreciate its formidability. With this knowledge, we can tune into what is inevitably part of our shared future. We can better position ourselves to stay informed about its developments. We can understand how our behavior informs the behavior of AI systems. And, as circumstances require, we can influence relevant policy.

I hope you have enjoyed this book, and I hope it has inspired you to further expand your understanding of artificial intelligence. We have only scratched the surface.

ACKNOWLEDGMENTS

The pages in this book were written in the early mornings of 2021 and 2022, before anyone in the house was awake. Sometimes, on weekends, those mornings would stretch into the late afternoon as I struggled to explain one concept or another. Little footsteps making their way up the stairs and into my office would interrupt my concentration. "Daddy, are you done work?" I would hear from my three-year-old daughter. Well, now I can say, yes, I am done, and thank you for being so wonderfully patient.

It is impossible to write a book, or produce anything, without having relied on others for help at some point. This work would not have been possible without the support of my wife, Maritza Marin, who encouraged me to try things I knew were too difficult when I was convinced I had no time to do them; thank you. Much is owed to Evan and Ellis Wenger, my wonderful and understanding children, who young as they are nevertheless sat beside me on many occasions, waiting in the agony of boredom as deep as only children can know but waiting still, until I finished a chapter or a paragraph, so we could go and play. My parents, Isaac Wenger and Barbara Luis, gambled their future to give me a better one. Thank you for your encouragement and for teaching me how to think.

This book benefited from the expertise of Ara Vartanian and Saumil Patel, who provided technical feedback that helped improve the readability and accuracy of the book in many ways. Thank you both for your contributions.

Two people at Working Fires Foundation carefully guided me through the entire publishing process. I know this book would have been deficient and less engaging without the expert touch of my editor, Matthew Perez, who painstakingly read and reread every single line, offering advice and adjusting the many mistakes I produced. This book is much better thanks to you. Any issues that remain can only be my fault. At the same time, Andrés Cruciani's contributions, help, and support made the publication and launching of this book considerably smoother and more enjoyable.

I have to express my gratitude to the many talented people involved in the production process: Fayyaz Ahmed (cover design), Wes Cowley and Marisa Crowley (proofreading), Joëlle O'Hanrahan (special edition slipcase design), James Protano (original book design), and Lisa Rivero (indexing).

To Damian Fozard and the entire CoreAVI team, thank you for your support and encouragement. Damian, thanks for your belief and for all the help you have given me in the production of this book, as well as for your support over our last decade working together.

For contributions that in one form or another have made this book possible, I must thank the following people: Amelia and Caridad Alfonso, Rocio Brito, Frank David de Castro, Pedro Carlos de Castro, Alex de la Cruz, Aidan Fabius, Dr. Farsad, Igor Krupin, Joe Liuzza and family, Carlos Luis, Dr. Sadeghian, Greg Szober, Kayvan Tirdad, Steve Viggers, and Dan Joncas.

This list ought to be much longer, and I am sure to have failed to remember key persons who have made a mark along the way or impressed upon me the need to write, and that I had some capacity to do so. To anyone I've omitted, whose valuable contributions live in the words set down here, I apologize and offer you my thanks.

And thank you again, dear reader, for reading.

SOURCES

The following works have inspired my writing.

Pale Blue Dot: A Vision of the Human Future in Space (Random House, 1994). Any scientist writing a popular science book owes a debt of gratitude to Carl Sagan, whether they realize it or not. Sagan was an American cosmologist and an incredible communicator, capable of explaining the most complex subjects in clear and accessible terms. He was also among the first scientists to advocate for informing the public of new developments in science. At the time, many of his peers heavily criticized his appearances on TV and his popular writings, as they considered a scientist's place to be the lab; according to them, anything else was a waste of time. Sagan understood that you don't have to be a scientist to appreciate science, and his efforts inspired millions, including myself.

The Selfish Gene (Oxford University Press, 1976). This book made Richard Dawkins one of the most important biologists of the twentieth century. *The Selfish Gene* is a groundbreaking piece that takes Darwin's theory of evolution by natural selection to a deeper level and identifies the fundamental actor driving evolution: the gene. Although it is purely a book about biology, I know it influenced my own writing.

The Organization of Behavior: A Neuropsychological Theory (John Wiley & Sons, 1949). Donald Hebb's seminal book on neural information processing was instrumental in developing the field of AI, and it directly informed the writing in this book by providing important background information and perspective on the development of neural networks, both biological and artificial.

Sapiens: A Brief History of Humankind (Harvill Secker / Signal, 2014) and
Homo Deus: A Brief History of Tomorrow (Harvill Secker, 2016). Yuval
Noah Harari's books on the history and future of humanity as critiques of
our social responsibilities provided invaluable inspiration.

The many lectures, research papers, and journal articles of Geoffrey Hinton,
Yann LeCun, Andrew Ng, and Yoshua Bengio contributed important
elements to all themes discussed in this book.

○ ● ○

Works whose influence can be found in specific chapters are
listed below.

Introduction: Living with Lions

Asher Hamilton, Isobel. "Elon Musk's Neuralink Wants to Embed Microchips
in People's Skulls and Get Robots to Perform Brain Surgery." Impact Lab,
Aug. 4, 2021. https://www.impactlab.com/2021/08/04/elon-musks-
neuralink-wants-to-embed-microchips-in-peoples-skulls-and-get-robots-to-
perform-brain-surgery/.

Reuters. "Maasai Now Track Lions instead of Killing Them." NBC News,
Oct. 15, 2009. https://www.nbcnews.com/id/wbna33240556.

Yasukawa, Olivia, and Thomas Page. "Lion-Killer Maasai Turn Wildlife
Warriors to Save Old Enemy." CNN, Feb. 8, 2017. https://www.cnn.
com/2017/02/07/africa/maasai-tanzania-wildlife-warriors/index.html.

Polarization and Its Consequences

Amari, S., and M. A. Arbib, eds. *Competition and Cooperation in Neural Nets:
Proceedings of the U.S.-Japan Joint Seminar Held at Kyoto, Japan February 15–
19, 1982.* Lecture Notes in Biomathematics Series. Berlin: Springer, 1982.
https://doi.org/10.1007/978-3-642-46466-9.

Burke, Robert E. "Sir Charles Sherrington's *The Integrative Action of the Nervous System*: A Centenary Appreciation." *Brain* 130, no. 4 (Apr. 2007): 887–94. https://doi.org/10.1093/brain/awm022.

Computer History Museum. "The Engines." The Babbage Engine. Accessed May 2, 2022. https://www.computerhistory.org/babbage/engines/.

Deng, J., W. Dong, R. Socher, L.-J. Li, Kai Li, and Li Fei-Fei. "ImageNet: A Large-Scale Hierarchical Image Database." In *2009 IEEE Conference on Computer Vision and Pattern Recognition*, 248–55. Washington, DC: IEEE Computer Society, 2009. https://doi.org/10.1109/CVPR.2009.5206848.

Gardner, Martin. "Mathematical Games: The Fantastic Combinations of John Conway's New Solitaire Game 'Life.'" *Scientific American* 223 (October 1970): 120–23. https://www.scientificamerican.com/magazine/sa/1970/10-01/.

Krizhevsky, Alex, Ilya Sutskever, and Geoffrey E. Hinton. "ImageNet Classification with Deep Convolutional Neural Networks." *Communications of the ACM* 60, no. 6 (June 2017): 84–90. https://doi.org/10.1145/3065386. Originally published in the proceedings of the 2012 Conference and Workshop on Neural Information Processing Systems, *Advances in Neural Information Processing Systems 25 (NIPS 2012)*, edited by F. Pereira, C. J. Burges, L. Bottou, and K. Q. Weinberger.

Large Movie Review Dataset. Accessed May 2, 2022. https://ai.stanford.edu/~amaas/data/sentiment/.

LeCun, Y., B. Boser, J. S. Denker, D. Henderson, R. E. Howard, W. Hubbard, and L. D. Jackel. "Backpropagation Applied to Handwritten Zip Code Recognition." Paper for AT&T Bell Laboratories, Holmdel, NJ, Sept. 1989. http://yann.lecun.com/exdb/publis/pdf/lecun-89e.pdf.

McCulloch, Warren, and Walter Pitts. "A Logical Calculus of Ideas Immanent in Nervous Activity." *Bulletin of Mathematical Biophysics* 5 (1943): 127–47. https://doi.org/10.1007/BF02478259.

Minsky, Marvin, and Seymour A. Papert. *Perceptrons: An Introduction to Computational Geometry*. Exp. ed. Cambridge, MA: MIT Press, 1987.

Ramón y Cajal, Santiago. *Texture of the Nervous System of Man and the Vertebrates*. Vol. 1. Translated and edited by Pedro Pasik and Tauba Pasik. Vienna: Springer, 1999.

Roberts, Eric. "History: The 1940's to the 1970's." Neural Networks (website). Accessed May 2, 2022. https://cs.stanford.edu/people/eroberts/courses/soco/projects/neural-networks/History/history1.html.

———. "History: The 1980's to the Present." Neural Networks (website). Accessed May 2, 2022. https://cs.stanford.edu/people/eroberts/courses/soco/projects/neural-networks/index.html.

Roberts, Siobhan. "The Lasting Lessons of John Conway's Game of Life." *New York Times*, Dec. 28, 2020. https://www.nytimes.com/2020/12/28/science/math-conway-game-of-life.html.

Sherrington, Charles Scott. *The Integrative Action of the Nervous System*. New Haven, CT: Yale University Press, 1906.

Swaine, M. R., and Paul A. Freiberger. "Analytical Engine." *Encyclopedia Britannica*, May 26, 2020. https://www.britannica.com/technology/Analytical-Engine.

Hello, Panda!

Elsayed, Gamaleldin, Shreya Shankar, Brian Cheung, Nicolas Papernot, Alex Kurakin, Ian Goodfellow, and Jascha Sohl-Dickstein. "Adversarial Examples that Fool both Computer Vision and Time-Limited Humans." arXiv.org, May 22, 2018. https://arxiv.org/abs/1802.08195.

Great Learning Team and Avinash Thite. "Introduction to VGG16: What Is VGG16?" *Great Learning*, Oct. 1, 2021. https://www.mygreatlearning.com/blog/introduction-to-vgg16/.

Harvard University. "A Nobel Partnership: Hubel & Wiesel." Harvard Brain
Tour. Accessed May 2, 2022. https://braintour.harvard.edu/archives/
portfolio-items/hubel-and-wiesel.

Hubel, David H., and Torsten N. Wiesel. "Early Exploration of the Visual
Cortex." *Neuron* 20 (March 1998): 401–12. https://doi.org/10.1016/
S0896-6273(00)80984-8.

Katz, Bernard. "Stephen William Kuffler, 24 August 1913–11 October
1980." *Biographical Memoirs of Fellows of the Royal Society* 28 (Nov.
1982): 225–59. https://doi.org/10.1098/rsbm.1982.0011.

Lian, Yanbo, Ali Almasi, David B. Grayden, Tatiana Kameneva, Anthony
N. Burkitt, and Hamish Meffin. "Learning Receptive Field Properties
of Complex Cells in V1." *PLOS Computational Biology*, Mar. 2, 2021.
https://doi.org/10.1371/journal.pcbi.1007957.

Nutan. "Deep Convolutional Networks VGG16 for Image Recognition
in Keras." Medium, Aug. 28, 2020. https://medium.com/@
nutanbhogendrasharma/deep-convolutional-networks-vgg16-for-image-
recognition-in-keras-a4beb59f80a7.

Simonyan, Karen, and Andrew Zisserman. "Very Deep Convolutional
Networks for Large-Scale Image Recognition." arXiv.org, Apr. 10, 2015.
https://arxiv.org/abs/1409.1556.

University of Bonn. "'Math Neurons' Identified in the Brain: When
Performing Calculations, Some Neurons Are Active when Adding,
Others when Subtracting." ScienceDaily, Feb. 14, 2022. https://www.
sciencedaily.com/releases/2022/02/220214121241.htm.

University of Oxford. Visual Geometry Group (homepage). Accessed May
2, 2022. https://www.robots.ox.ac.uk/~vgg/.

Answering an Age-Old Question

Asimov, Isaac. "Runaround." In *I, Robot*, 25–45. New York: Bantam, 1991.

Bryson, Arthur E., Jr., and Yu-Chi Ho. *Applied Optimal Control: Optimization, Estimation, and Control*. New York: Taylor & Francis, 1975.

Hintze, Arend. "What an Artificial Intelligence Researcher Fears about AI." *Scientific American*, July 14, 2017. https://www.scientificamerican.com/article/what-an-artificial-intelligence-researcher-fears-about-ai/.

LeCun, Y. "A Theoretical Framework for Back-Propagation." In *Proceedings of the 1988 Connectionist Models Summer School*, edited by D. Touretzky, G. Hinton, and T. Sejnowski, 21–28. Pittsburgh: Carnegie Mellon University, 1988. http://yann.lecun.com/exdb/publis/pdf/lecun-88.pdf.

Piper, Kelsey. "Why Elon Musk Fears Artificial Intelligence." Vox, Nov. 2, 2018. https://www.vox.com/future-perfect/2018/11/2/18053418/elon-musk-artificial-intelligence-google-deepmind-openai.

Russell, Stuart, and Peter Norvig, eds. *Artificial Intelligence: A Modern Approach*. 4th ed. Hoboken, NJ: Pearson, 2021.

Shead, Sam. "Elon Musk Says DeepMind Is His 'Top Concern' when It Comes to A.I." CNBC, July 29, 2020. https://www.cnbc.com/2020/07/29/elon-musk-deepmind-ai.html.

Intelligent Discourse

Agence-France Presse in Shanghai. "World's Best Go Player Flummoxed by Google's 'Godlike' AlphaGo AI." *The Guardian*, May 23, 2017. https://www.theguardian.com/technology/2017/may/23/alphago-google-ai-beats-ke-jie-china-go.

"AlphaGo—the Movie: Full Award-Winning Documentary." YouTube video posted by DeepMind on Mar. 13, 2020, 1:30:27. https://www.youtube.com/watch?v=WXuK6gekU1Y.

Angwin, Julia, Jeff Larson, Surya Mattu, and Lauren Kirchner. "Courtroom Equations Wrongly Flagging Blacks as Future Criminals." *Tampa Bay Times*, May 23, 2016. https://www.tampabay.com/news/publicsafety/crime/courtroom-equations-wrongly-flagging-blacks-as-future-criminals/2278656/.

BBC. "Google AI Defeats Human Go Champion." BBC News, May 25, 2017. https://www.bbc.com/news/technology-40042581.

Bostrom, Nick. "The Superintelligent Will: Motivation and Instrumental Rationality in Advanced Artificial Agents." *Minds and Machines* 22, no. (May 2012): 71–85. https://doi.org/10.1007/s11023-012-9281-3.

Cummings, M. L. "Automation Bias in Intelligent Time Critical Decision Support Systems." AIAA Meeting Paper, AIAA 1st Intelligent Systems Technical Conference, Sept. 20–22, 2004, Chicago. https://doi.org/10.2514/6.2004-6313.

Dressel, Julia, and Hany Farid. "The Accuracy, Fairness, and Limits of Predicting Recidivism." *Science Advances* 4, no. 1 (Jan. 2018). https://doi.org/10.1126/sciadv.aao5580.

Gorner, Jeremy. "Chicago Police Use 'Heat List' as Strategy to Prevent Violence." *Chicago Tribune*, Aug. 21, 2013. https://www.chicagotribune.com/news/ct-xpm-2013-08-21-ct-met-heat-list-20130821-story.html.

Lin, Zhiyuan "Jerry," Jongbin Jung, Sharad Goel, and Jennifer Skeem. "The Limits of Human Predictions of Recidivism." *Science Advances* 6, no. 7 (Feb. 2020). https://doi.org/10.1126/sciadv.aaz0652.

Russell, Stuart. *Human Compatible: Artificial Intelligence and the Problem of Control*. Repr. ed. New York: Penguin Books, 2020.

Silver, David, Thomas Hubert, Julian Schrittwieser, Ioannis Antonoglou, Matthew Lai, Arthur Guez, Marc Lanctot, Laurent Sifre, Dharshan Kumaran, Thore Graepel, Timothy Lillicrap, Karen Simonyan, and Demis Hassabis. "Mastering Chess and Shogi by Self-Play with a General Reinforcement Learning Algorithm." arXiv.org, Dec. 5, 2017. https://arxiv.org/abs/1712.01815.

Silver, David, Thomas Hubert, Julian Schrittwieser, and Demis Hassabis. "AlphaZero: Shedding New Light on Chess, Shogi, and Go." DeepMind, Dec. 6, 2018. https://www.deepmind.com/blog/alphazero-shedding-new-light-on-chess-shogi-and-go.

Simonite, Tom. "Algorithms Were Supposed to Fix the Bail System. They Haven't." Wired, Feb. 19, 2020. https://www.wired.com/story/algorithms-supposed-fix-bail-system-they-havent/.

Stroud, Matt. "Heat Listed." The Verge, May 24, 2021. https://www.theverge.com/c/22444020/chicago-pd-predictive-policing-heat-list.

U.S. Department of Health and Human Services. "Heart Disease and African Americans." Office of Minority Health, last modified Jan. 1, 2022. https://minorityhealth.hhs.gov/omh/browse.aspx?lvl=4&lvlid=19#:~:text=In%202018%2C%20African%20Americans%20were,their%20blood%20pressure%20under%20control.

Yudkowsky, Eliezer. "MIRI Announces New 'Death with Dignity' Strategy." LessWrong, Apr. 1, 2022. https://www.lesswrong.com/posts/j9Q8bRmwCgXRYAgcJ/miri-announces-new-death-with-dignity-strategy.

Yudkowsky, Eliezer, Anna Salamon, Carl Shulman, Steven Kaas, Tom McCabe, and Rolf Nelson. Reducing Long-Term Catastrophic Risks from Artificial Intelligence. San Francisco: Singularity Institute, 2010. https://intelligence.org/files/ReducingRisks.pdf.

Završnik, Aleš. "Criminal Justice, Artificial Intelligence Systems, and Human Rights." ERA Forum 20 (2020): 567–83. https://doi.org/10.1007/s12027-020-00602-0.

From Chat Rooms to Chatbots

Bonifacic, I. "Alphabet's Isomorphic Labs Is a New Company Focused on AI-Driven Drug Discovery." Engadget, Nov. 4, 2021. https://www.engadget.com/alphabet-isomorphic-labs-announcement-193546886.html.

Coldewey, Devin. "Isomorphic Labs Is Alphabet's Play in AI Drug Discovery." Tech Crunch, Nov. 4, 2021. https://techcrunch.com/2021/11/04/isomorphic-labs-is-alphabets-play-in-ai-drug-discovery.

INDEX

Page numbers in italics refer to figures and tables.